# Options and Derivatives Programming in C++

Algorithms and Programming Techniques for the Financial Industry

Carlos Oliveira

Apress®

*Options and Derivatives Programming in C++*

Carlos Oliveira
Monmouth Junction, New Jersey
USA

ISBN-13 (pbk): 978-1-4842-1813-6     ISBN-13 (electronic): 978-1-4842-1814-3
DOI 10.1007/978-1-4842-1814-3

Library of Congress Control Number: 2016954432

Copyright © 2016 by Carlos Oliveira

This work is subject to copyright. All rights are reserved by the Publisher, whether the whole or part of the material is concerned, specifically the rights of translation, reprinting, reuse of illustrations, recitation, broadcasting, reproduction on microfilms or in any other physical way, and transmission or information storage and retrieval, electronic adaptation, computer software, or by similar or dissimilar methodology now known or hereafter developed.

Trademarked names, logos, and images may appear in this book. Rather than use a trademark symbol with every occurrence of a trademarked name, logo, or image we use the names, logos, and images only in an editorial fashion and to the benefit of the trademark owner, with no intention of infringement of the trademark.

The use in this publication of trade names, trademarks, service marks, and similar terms, even if they are not identified as such, is not to be taken as an expression of opinion as to whether or not they are subject to proprietary rights.

While the advice and information in this book are believed to be true and accurate at the date of publication, neither the authors nor the editors nor the publisher can accept any legal responsibility for any errors or omissions that may be made. The publisher makes no warranty, express or implied, with respect to the material contained herein.

> Managing Director: Welmoed Spahr
> Lead Editor: Jonathan Gennick
> Technical Reviewer: Don Reamey
> Editorial Board: Steve Anglin, Pramila Balan, Laura Berendson, Aaron Black, Louise Corrigan, Jonathan Gennick, Todd Green, Robert Hutchinson, Celestin Suresh John, Nikhil Karkal, James Markham, Susan McDermott, Matthew Moodie, Natalie Pao, Gwenan Spearing
> Coordinating Editor: Jill Balzano
> Copy Editor: Kezia Endsley
> Compositor: SPi Global
> Indexer: SPi Global
> Artist: SPi Global
> Cover Designer: Anna Ishchenko

Distributed to the book trade worldwide by Springer Science+Business Media New York, 233 Spring Street, 6th Floor, New York, NY 10013. Phone 1-800-SPRINGER, fax (201) 348-4505, e-mail orders-ny@springer-sbm.com, or visit www.springer.com. Apress Media, LLC is a California LLC and the sole member (owner) is Springer Science + Business Media Finance Inc (SSBM Finance Inc). SSBM Finance Inc is a Delaware corporation.

For information on translations, please e-mail rights@apress.com, or visit www.apress.com.

Apress and friends of ED books may be purchased in bulk for academic, corporate, or promotional use. eBook versions and licenses are also available for most titles. For more information, reference our Special Bulk Sales–eBook Licensing web page at www.apress.com/bulk-sales.

Any source code or other supplementary material referenced by the author in this text is available to readers at www.apress.com. For detailed information about how to locate your book's source code, go to www.apress.com/source-code/.

*To my family, my real source of inspiration.*

# Contents at a Glance

About the Author .................................................................................................. xiii
About the Technical Reviewer ............................................................................... xv
Introduction ........................................................................................................ xvii

■Chapter 1: Options Concepts ............................................................................... 1
■Chapter 2: Financial Derivatives ........................................................................ 19
■Chapter 3: Basic Algorithms .............................................................................. 35
■Chapter 4: Object-Oriented Techniques ............................................................. 67
■Chapter 5: Design Patterns for Options Processing ........................................... 85
■Chapter 6: Template-Based Techniques ........................................................... 101
■Chapter 7: STL for Derivatives Programming .................................................. 115
■Chapter 8: Functional Programming Techniques ............................................. 127
■Chapter 9: Linear Algebra Algorithms .............................................................. 143
■Chapter 10: Algorithms for Numerical Analysis ............................................... 161
■Chapter 11: Models Based on Differential Equations ...................................... 175
■Chapter 12: Basic Models for Options Pricing .................................................. 189
■Chapter 13: Monte Carlo Methods .................................................................... 207
■Chapter 14: Using C++ Libraries for Finance ................................................... 223
■Chapter 15: Credit Derivatives ......................................................................... 241

Index ................................................................................................................. 255

# Contents

About the Author ............................................................................................ xiii

About the Technical Reviewer ........................................................................ xv

Introduction .................................................................................................. xvii

## ■Chapter 1: Options Concepts ........................................................................ 1

Basic Definitions ..................................................................................................... 2

Option Greeks .......................................................................................................... 6

Using C++ for Options Programming ....................................................................... 8

    Availability ........................................................................................................................ 8

    Performance .................................................................................................................... 9

    Standardization ............................................................................................................... 9

    Expressiveness ............................................................................................................. 10

Modeling Options in C++ ....................................................................................... 10

    Creating Well-Behaving Classes ................................................................................... 10

    Computing Option Value at Expiration .......................................................................... 12

    Complete Listing ........................................................................................................... 13

    Building and Testing ..................................................................................................... 16

Further References ................................................................................................ 18

Conclusion ............................................................................................................. 18

## ■Chapter 2: Financial Derivatives ................................................................. 19

Models for Derivative Pricing ................................................................................ 19

    Credit Default Swaps .................................................................................................... 21

    Collateralized Debt Obligations .................................................................................... 22

FX Derivatives ........................................................................................................23
Derivative Modeling Equations ............................................................................23
Numerical Models .................................................................................................24
Binomial Trees ......................................................................................................24
Simulation Models ................................................................................................25

## Using the STL ..........................................................................................26
Generating a Random Walk ..................................................................................27
Complete Listing ...................................................................................................29
Building and Testing .............................................................................................32

## Further References ................................................................................33

## Conclusion .............................................................................................34

# Chapter 3: Basic Algorithms ..................................................................35
## Date and Time Handling .......................................................................35
Date Operations ....................................................................................................36
Complete Listings .................................................................................................39

## A Compact Date Representation ..........................................................48
Complete Listings .................................................................................................49
Building and Testing .............................................................................................53

## Working with Networks ........................................................................53
Creating a Dictionary Class ..................................................................................54
Calculating a Shortest Path ..................................................................................56
Complete Listings .................................................................................................58
Building and Testing .............................................................................................66

## Conclusion .............................................................................................66

# Chapter 4: Object-Oriented Techniques ................................................67
## OO Programming Concepts ..................................................................67
Encapsulation ........................................................................................................69
Inheritance ............................................................................................................72
Polymorphism .......................................................................................................72
Polymorphism and Virtual Tables .........................................................................75

Virtual Functions and Virtual Destructors .................................................................. 76
Abstract Functions ........................................................................................................ 78
Building Class Hierarchies ........................................................................................... 80
Object Composition ...................................................................................................... 82

Conclusion ........................................................................................................................... 83

# Chapter 5: Design Patterns for Options Processing ............................................. 85

Introduction to Design Patterns ........................................................................................... 86

The Factory Method Design Pattern .................................................................................... 87

The Singleton Pattern ........................................................................................................... 90
Clearing House Implementation in C++ ...................................................................... 91

The Observer Design Pattern ............................................................................................... 93
Complete Code ............................................................................................................. 96

Conclusion ........................................................................................................................... 100

# Chapter 6: Template-Based Techniques ............................................................. 101

Introduction to Templates ................................................................................................... 101

Compilation-Time Polymorphism ....................................................................................... 103

Template Functions ............................................................................................................. 104

Implementing Recursive Functions .................................................................................... 106

Recursive Functions and Template Classes ....................................................................... 108

Containers and Smart Pointers ........................................................................................... 109

Avoiding Lengthy Template Instantiations ......................................................................... 111
Pre-Instantiating Templates ........................................................................................ 111

Conclusion ........................................................................................................................... 113

# Chapter 7: STL for Derivatives Programming ..................................................... 115

Introduction to Algorithms in the STL ................................................................................ 115

Sorting .................................................................................................................................. 116
Presenting Frequency Data ......................................................................................... 119

Copying Container Data ..................................................................................................... 121

Finding Elements ................................................................................................................. 123

ix

Selecting Option Data ................................................................................................... 124
Conclusion ................................................................................................................ 126

## Chapter 8: Functional Programming Techniques ................................................. 127

Functional Programming Concepts .............................................................................. 128
Function Objects ........................................................................................................ 128
Functional Predicates in the STL ................................................................................ 131
The Bind Function ..................................................................................................... 133
Lambda Functions in C++11 ..................................................................................... 135
Complete Code .......................................................................................................... 137
Conclusion ................................................................................................................ 142

## Chapter 9: Linear Algebra Algorithms ................................................................... 143

Vector Operations ...................................................................................................... 143
    Scalar-to-Vector Operations ................................................................................. 144
    Vector-to-Vector Operations ................................................................................ 146
Matrix Implementation .............................................................................................. 148
Using the uBLAS Library ........................................................................................... 154
Complete Code .......................................................................................................... 156
Conclusion ................................................................................................................ 160

## Chapter 10: Algorithms for Numerical Analysis .................................................... 161

Representing Mathematical Functions ........................................................................ 161
    Using Horner's Method ........................................................................................ 163
Finding Roots of Equations ........................................................................................ 165
    Newton's Method ................................................................................................ 165
Integration ................................................................................................................ 170
Conclusion ................................................................................................................ 173

## Chapter 11: Models Based on Differential Equations ........................................... 175

General Differential Equations ................................................................................... 175
Ordinary Differential Equations ................................................................................. 176

| Euler's Method | 177 |
| Implementing the Method | 178 |
| The Runge-Kutta Method | 182 |
| Runge-Kutta Implementation | 183 |
| Complete Code | 185 |
| Conclusion | 187 |

## Chapter 12: Basic Models for Options Pricing .................................. 189

| Lattice Models | 189 |
| Binomial Model | 190 |
| Binomial Model Implementation | 192 |
| Pricing American-Style Options | 195 |
| Solving the Black-Scholes Model | 197 |
| Numerical Solution of the Model | 198 |
| Complete Code | 202 |
| Conclusion | 205 |

## Chapter 13: Monte Carlo Methods ............................................... 207

| Introduction to Monte Carlo Methods | 207 |
| Random Number Generation | 208 |
| Probability Distributions | 210 |
| Using Common Probability Distributions | 213 |
| Creating Random Walks | 218 |
| Conclusion | 221 |

## Chapter 14: Using C++ Libraries for Finance ................................... 223

| Boost Libraries | 223 |
| Installing Boost | 225 |
| Solving ODEs with Boost | 225 |
| Solving a Simple ODE | 227 |

## The QuantLib Library .................................................................................................... 229

### Handling Dates ............................................................................................... 230
### Working with Calendars.................................................................................. 231
### Computing Solutions for Black-Scholes Equations ............................................ 233
### Creating a C++ Interface................................................................................. 235
### Complete Code .............................................................................................. 237

## Conclusion ........................................................................................................ 240

# Chapter 15: Credit Derivatives ............................................................................ 241

## Introduction to Credit Derivatives ...................................................................... 241
## Modeling Credit Derivatives ............................................................................. 242
## Using Barrier Options ....................................................................................... 243

### A Solver Class for Barrier Options ..................................................................... 244
### Barrier Option Classes in QuantLib.................................................................... 245
### An Example Using QuantLib............................................................................. 247

## Complete Code ................................................................................................ 249
## Conclusion ...................................................................................................... 253

# Index................................................................................................................... 255

# About the Author

**Carlos Oliveira** works in the area of quantitative finance, with more than 10 years of experience in creating scientific and financial models in C++. During his career, Carlos has developed several large-scale applications for financial companies such as Bloomberg L.P. and Incapital LLC. Oliveira obtained a PhD in Operations Research and Systems Engineering from the University of Florida, an MSc in Computer Science from UFC (Brazil), and a BSc in Computer Science from UECE (Brazil). He has also performs academic research in the field of combinatorial optimization, with applications in diverse areas such as finance, telecommunications, computational biology, and logistics. Oliveira has written more than 30 academic papers on optimization and authored three books, including Practical C++ Financial Programming (Apress, 2015).

# About the Technical Reviewer

 **Don Reamey** is currently a Senior Engineering Manager at Apptio Corporation. Previously he worked at Microsoft as a Principal Software Engineer and Program Manager. Don also managed engineering teams at TIBCO Software and Bank of America. He enjoys learning and creating new programming languages, creating web-based applications, and playing guitar and weight lifting.

# Introduction

On Wall Street, the use of algorithmic trading and other computational techniques has skyrocketed in the last few years, as can be seen from the public interest in automated trading as well as the profits generated by these strategies. This growing trend demonstrates the importance of using software to analyze and trade markets in diverse areas of finance. One particular area that has been growing in importance during the last decade is options and derivatives trading.

Initially used only as a niche investment strategy, derivatives have become one of the most common instruments for investors in all areas. Likewise, the interest in automated trading and analysis of such instruments has also increased considerably.

Along with scientists and economists, software engineers have also greatly contributed to the development of advanced computational techniques using financial derivatives. Such techniques have been used at banks, hedge funds, pension funds, and other financial institutions. In fact, every day new systems are developed to give a trading advantage to the players in this industry.

This books attempts to provide the basic programming knowledge needed by C++ programmers working with options and derivatives in the financial industry. This is a hands-on book for programmers who want to learn how C++ is used to develop solutions for options and derivatives trading. In the book's chapters, you'll explore the main algorithms and programming techniques used in the implementation of systems and solutions for trading options and other derivatives.

Because of stringent performance characteristics, most of these trading systems are developed using C++ as the main implementation language. This makes the topic of this book relevant to everyone interested in programming skills used in the financial industry in general.

In Options and Derivatives Programming in C++, I cover the features of the language that are more frequently used to write financial software for options and derivatives. These features include the STL, templates, functional programming, and support for numerical libraries. New features introduced in the latest updates of the C++ standard are also covered, including additional functional techniques such as lambda functions, automatic type detection, custom literals, and improved initialization strategies for C++ objects.

I also provide how-to examples that cover the major tools and concepts used to build working solutions for quantitative finance. The book teaches you how to employ advanced C++ concepts as well as the basic building libraries used by modern C++ developers, such as the STL, Boost, and QuantLib. It also discusses how to create correct and efficient applications, leveraging knowledge of object-oriented and template-based programming. I assume only a basic knowledge of C and C++ and extensively use concepts already mastered by developers who are fluent in these languages.

In the process of writing this book, I was also concerned with providing value for readers who are trying to use their current programming knowledge in order to become proficient at the style of programming used in large banks, hedge funds, and other investment institutions. Therefore, the topics covered in the book are introduced in a logical and structured way. Even novice programmers will be able to absorb the most important topics and competencies necessary to develop C++ for the problems occurring on the analysis of options and other financial derivatives.

■ INTRODUCTION

# Audience

This book is intended for readers who already have a working knowledge of programing in C, C++, or another mainstream language. These are usually professionals or advanced students in computer science, engineering, and mathematics, who have interest in learning about options and derivatives programming using the C++ language, for personal or for professional reasons. The book is also directed at practitioners of C++ programming in financial institutions, who would use the book as a ready-to-use reference of software development algorithms and best practices for this important area of finance.

Many readers are interested in a book that would describe how modern C++ techniques are used to solve practical problems arising when considering options on financial instruments and other derivatives. Being a multi-paradigm language, C++ usage may be slightly different in each area, so the skills that are useful for developing desktop applications, for example, are not necessarily the same ones used to write high-performance software.

A large part of high-performance financial applications are written in C++, which means that programmers who want to enter this lucrative market need to acquire a working knowledge of specific parts of the language. This book attempts to give developers who want to develop their knowledge effectively this choice, while learning one of the most sought-after and marketable skillsets for modern financial application and high-performance software development.

This book is also targeted at students and new developers who have some experience with the C++ language and want to leverage that knowledge into financial software development. This book is written with the goal of reaching readers who need a concise, algorithms-based strategy, providing basic information through well-targeted examples and ready-to-use solutions. Readers will be able to directly apply the concepts and sample code to some of the most common problems faced regarding the analysis of options and derivative contracts.

# What You Will Learn

Here is a sample of topics that are covered in the following chapters:

- Fundamental problems in the options and derivatives market
    - Options market models
    - Derivative valuation problems
    - Trading strategies for options and derivatives
- Pricing algorithms for derivatives
    - Binomial method
    - Differential equations method
    - Black-Scholes model
- Quantitative finance algorithms for options and derivatives
    - Linear algebra techniques
    - Interpolation
    - Calculating roots
    - Numerical solution for PDEs

- Important features of C++ language as used in quantitative financial programming, such as
    - Templates
    - STL containers
    - STL algorithms
    - Boost libraries
    - QuantLib
    - New features of C++11 and C++14

## Book Contents

Here is a quick overview of the major topics covered in each chapter.

- Chapter 1: "Options Concepts." An option is a standard financial contract that derives its value from an underlying asset such as a stock. Options can be used to pursue multiple economic objectives, such as hedging against variations on the underlying asset, or speculating on the future price of a stock. Chapter 1 presents the basic concepts of options, including their advantages and challenges. It also explains how options can be modeled using C++. The main topics covered in this chapter are as follows:
    - Basic definitions of options
    - An introduction to options strategies
    - Describing options with Greeks
    - Sample code for options handling

- Chapter 2: "Financial Derivatives." A derivative is a general term for a contract whose price is based on an underlying asset. In the previous decades, the financial industry created and popularized a large number of derivatives. Pricing and trading these derivatives is a large part of the work performed by trading desks throughout the world. Derivatives have been created based on diverse assets such as foreign currency, mortgage contracts, and credit default swaps. This chapter explores this type of financial instrument and presents a few C++ techniques to model specific derivatives. The main topics covered in this chapter are as follows:
    - Credit default swaps
    - Forex derivatives
    - Interest rate derivatives
    - Exotic derivatives

## INTRODUCTION

- Chapter 3: "Basic Algorithms." To become a proficient C++ developer, it is essential to have good knowledge of the basic algorithms used in your application area. Some basic algorithms for tasks such as vector processing, date and time handling, and data access and storage are useful in almost all applications involving options and other financial derivatives. This chapter surveys such algorithms and their implementation in C++, including the following topics:
    - Date and time handling
    - Vector processing
    - Graphs and networks
    - Fast data processing
- Chapter 4: "Object-Oriented Techniques." For the last 30 years, object-oriented techniques have become the standard for software development. Since C++ fully supports OO programming, it is imperative that you have a good understanding of OO techniques in order to solve the problems presented by options and derivatives. I present summary of what you need to become proficient in the relevant OO techniques used in the financial industry. Some of the topics covered in this chapter are:
    - Problem partitioning
    - Designing solutions using OO strategies
    - OO implementation in C++
    - Reusing OO components
- Chapter 5: "Design Patterns for Options Processing." Design patterns are a set of common programming practices that can be used to simplify the solution of recurring problems. With the use of OO techniques, design patterns can be cleanly implemented as a set of classes that interact toward the solution of a common goal. In this chapter, you learn about the most common design pattern employed when working with financial options and derivatives, with specific examples. It covers the following topics:
    - The importance of design patterns
    - Factory pattern
    - Visitor pattern
    - Singleton pattern
    - Less common patterns
- Chapter 6: "Template-Based Techniques." C++ templates allow programmers to write code that works without modification on different data types. Through the careful use of templates, C++ programmers can write code with high performance and low overhead, without the need to employ more computationally expensive object-oriented techniques. This chapter explores a few template-oriented practices used in the solution of options and derivatives-based financial problems.
    - Motivating the use of templates
    - Compile-time algorithms
    - Containers and smart pointers
    - Template libraries

- Chapter 7: "STL for Derivatives Programming." Modern financial programming in C++ makes heavy use of template-based algorithms. Many of the basic template algorithms are implemented in the standard template library (STL). This chapter discusses the STL and how it can be used in quantitative finance projects, in particular to solve options and financial derivative problems. You will get a clear understanding of how the STL interacts with other parts of the system, and how it imposes a certain structure on classes developed in C++.
    - STL-based algorithms
    - Functional techniques on STL
    - STL containers
    - Smart pointers
- Chapter 8: "Functional Programming Techniques." Functional programming is a technique that focuses on the direct use of functions as first-class objects. This means that you are allowed to create, store, and call functions as if they were just another variable of the system. Recently, functional techniques in C++ have been greatly improved with the adoption of the new standard (C++11), particularly with the introduction of lambda functions. The following topics are explored in this chapter:
    - Lambdas
    - Functional templates
    - Functions as first-class objects
    - Managing state in functional programming
    - Functional techniques for options processing
- Chapter 9: "Linear Algebra Algorithms." Linear algebra techniques are used throughout the area of financial engineering, and in particular in the analysis of options and other financial derivatives. Therefore, it is important to understand how the traditional methods of linear algebra can be applied in C++. With this goal in mind, I present a few examples that illustrate how to use some of the most common linear algebra algorithms. In this chapter, you will also learn how to integrate existing LA libraries into your code.
    - Implementing matrices
    - Matrix decomposition
    - Computing determinants
    - Solving linear systems of equations

# INTRODUCTION

- Chapter 10: "Algorithms for Numerical Analysis." Equations are some of the building blocks of financial algorithms for options and financial derivatives, and it is important to be able to efficiently calculate the solution for such mathematical models. In this chapter, you will see programming recipes for different methods of calculating equation roots and integrating functions, along with explanations of how they work and when they should be used. I also discuss numerical error and stability issues that present a challenge for developers in the area of quantitative financial programming.
    - Basic numerical algorithms
    - Root-finding algorithms
    - Integration algorithms
    - Reducing errors in numeric algorithms
- Chapter 11: "Models Based on Differential Equations." Differential equations are at the heart of many techniques using in the analysis of derivatives. There are several processes for solving and analyzing PDEs that can be implemented in C++. This chapter presents programming recipes that cover aspects of PDE-based option modeling and application in C++. Topics covered include the following:
    - Basic techniques for differential equations
    - Ordinary differential equations
    - Partial difference equations
    - Numerical algorithms for differential equations
- Chapter 12: "Basic Models for Options Pricing." Options pricing is the task of determining the fair value of a particular option, given a set of parameters that exactly determine the option type. This chapter discusses some of the most popular models for options pricing. They include tree-based methods, such as binomial and trinomial trees. It also discusses the famous Black-Scholes model, which is frequently used as the basis for the analysis of most options and derivative contracts.
    - Binomial trees
    - Trinomial trees
    - Black-Scholes model
    - Implementation strategies
- Chapter 13: "Monte Carlo Methods." Among other programming techniques for equity markets, Monte Carlo simulation has a special place due to its wide applicability and easy implementation. These methods can be used to forecast prices or to validate options buying strategies, for example. In This chapter provides programming recipes that can be used as part of simulation-based algorithms applied to options pricing.
    - Probability distributions
    - Random number generation
    - Stochastic models for options
    - Random walks
    - Improving performance

- Chapter 14: "Using C++ Libraries for Finance." Writing good financial code is not an individual task. You frequently have to use libraries created by other developers and integrate them into your own work. In the world of quantitative finance, a number of C++ libraries have been used with great success. This chapter reviews some of these libraries and explains how they can be integrated into your own derivative-based applications. Some of the topics covered include the following:
  - Standard library tools
  - QuantLib
  - Boost math
  - Boost lambda

- Chapter 15: "Credit Derivatives." Credit derivatives are an increasingly popular type of financial derivative that aims at reducing credit risk—that is, the risk of default posed by contracts established with a counterparty. Credit derivatives can be modeled using some of the tools already discussed for options, although credit derivative have their own peculiarities. This chapter describes how to create the C++ code needed to quantitatively analyze such financial contracts. Here are some of the topics discussed:
  - General concepts of credit derivatives
  - Modeling the problem
  - C++ algorithms for derivative pricing
  - Improving algorithm efficiency

## Example Code

The examples given in this book have all been tested on MacOS X using the Xcode 7 IDE. The code uses only standard C++ techniques, so you should be able to build the given examples using any standards-compliant C++ compiler that implements the C++11 standard. For example, gcc is available on most platforms, and Microsoft Visual Studio will also work on Windows.

If you use MacOS X and don't have Xcode installed in your computer yet, you can download it for free from Apple's developer web site at http://developer.apple.com.

If you instead prefer to use MinGW on Windows, you can download the MinGW distribution from the web site http://www.mingw.org.

Once MinGW is installed, start the command prompt from the MinGW program group in the Start menu. Then, you can type gcc to check that the compiler is properly installed.

To download the source code for all examples in this book, visit the web page of the author at http://coliveira.net.

# CHAPTER 1

# Options Concepts

In the last few decades, software development has become an integral part of the investment industry. Advances in trading infrastructure, as well as the need for increased volume and liquidity, has caused financial institutions to adopt computational techniques in their day-to-day operations. This means that there are many opportunities for computer scientists specializing in the design and development of automated strategies for trading and analyzing options and other financial derivatives.

Options are among the several investment vehicles that are currently traded using automated methods, as you will learn in the following chapters. Given the mathematical structure of options and related derivatives, it is possible to explore their features in a controlled way, which is ideal for the application of computational algorithms. In this book, I present many of the computational techniques used to develop strategies in order to trade options and other derivatives.

An *option* is a standard financial contract that derives its value from an underlying asset such as common stock or commodities. Options can be used to pursue multiple economic objectives, such as hedging against large variations on the underlying asset, or speculating on the future price of a stock. This chapter presents the basic concepts of options, along with supporting definitions. I also give an overview of the use of C++ in the financial industry, and how options can be modeled using C++.

The following concepts are explored in the next sections:

- *Basic definitions:* You will learn fundamental definitions about options contracts and how they are used in the investment industry.

- *Fundamental option strategies:* Due to their flexibility, options can be combined in a surprising large number of investment strategies. You will learn about some of the most common option strategies, and how to model them using C++.

- *Option Greeks:* One of the advantages of options investing is that it promotes a very analytical view of financial decisions. Each option is defined by a set of variables called *Greeks*, which reflect the properties of an option contract at each moment in time.

- *Delta hedging:* One of the ways to use options is to create a hedge for some other underlying asset positions. This is called delta hedging, and it is widely used in the financial industry. You will see how this investment technique works and how it can be modeled using C++.

---

**Electronic supplementary material** The online version of this chapter (doi:10.1007/978-1-4842-1814-3_1) contains supplementary material, which is available to authorized users.

© Carlos Oliveira 2016
C. Oliveira, *Options and Derivatives Programming in C++*, DOI 10.1007/978-1-4842-1814-3_1

CHAPTER 1 ▪ OPTIONS CONCEPTS

# Basic Definitions

Let's start with an overview of concepts and programming problems presented by options in the financial industry. Options are specialized trading instruments, and therefore require their users to be familiar with a number of details about their operation. In this section, I introduce some basic definitions about options and their associated ideas. Before starting, take a quick look at Table 1-1 for a summary of the most commonly used concepts. These concepts are defined in more detail in the remaining parts of this section.

***Table 1-1.*** *Basic Concepts in Options Trading*

| Concept | Definition |
| --- | --- |
| Call | An option contract that gives its owner the right to buy the underlying asset for a predetermined price. |
| Put | An option contract that gives its owner the right to sell the underlying asset for a predetermined price. |
| Underlying | Asset whose price is used as the base of the options contract. |
| Strike price | The price at which option owners can buy or sell the underlying asset under the options contract. |
| Expiration | The last date of the options contract. |
| Settlement | The act of exercising the options contract at the expiration date. |
| Intrinsic value | Amount of option value that is directly derived from the underlying price. |
| Break-even price | The price at which an investor will start to make a profit in the option. |
| Exercise | The act of buying of selling the option underlying under the price determined by the option contract. |
| American option | An option style where option owners can exercise the option at any moment between option purchase and option expiration. |
| European option | An option style where option owners can exercise the option only at expiration time. |
| ATM | (At The Money): Refers to options that have a strike price close to the current price for the underlying. |
| OTM | (Out of The Money): Refers to options that have a strike price above (for calls) or below (for puts) the current price of the underlying asset. These options have no intrinsic value. |
| ITM | (In The Money): Refers to options that have a strike price below (for calls) or above (for puts) the current price of the underlying asset. These options have an intrinsic value. |

Options can be classified according to several criteria. The features of the options determine every aspect of how they can be used, such as the quantity of underlying assets, the strike price, and the expiration, among others. There are two main types of Options processing: calls and puts. A *call* is a standard contract that gives its owner the right (but not the obligation) to buy an underlying instrument at a particular price. Similarly, a *put* is standard contract that gives its owner the right (but not the obligation) to sell an underlying instrument at a predetermined price.

The *strike price* is the price at which the option can be exercised. For example, a call for IBM stock with strike $100 gives its owner the right to buy IBM stock at the price of $100. If the current price of IBM is greater than $100, the owner of such an option has the right to buy the stock at a price that is lower than the current price, which means that the call has a higher value as the value of IBM stock increases. This situation is exemplified in Figure 1-1. If the current price is lower than $100 at expiration, the value of the option is zero, since there is no profit in exercising the contract.

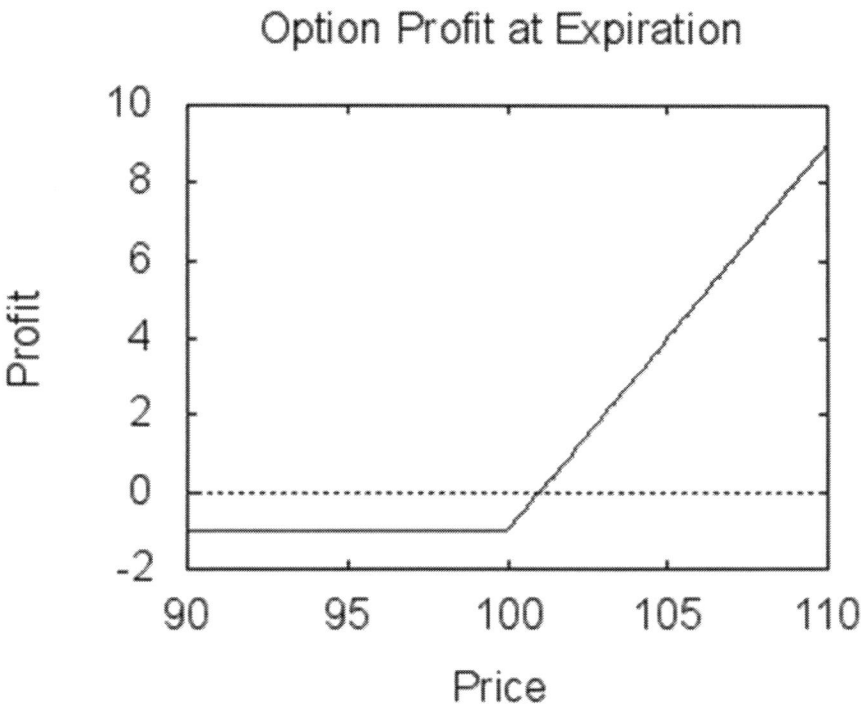

*Figure 1-1. Profit chart for a call option*

As you have seen in this example, if you buy an option, you have an unlimited gain and your losses are limited to the value paid originally. This is advantageous when you're trying to limit losses in a particular investment scenario. As long as you are okay with losing a limited amount of money paid for the option, you can profit from the unlimited upside potential of a call (if the underlying grows in price). Put options don't have unlimited profit potential since the maximum gain happens when the underlying price is zero. However, they still benefit from the well-defined, limited loss versus the possible large gains.

*Expiration*: The expiration is the moment when the option contract ends its validity and a final value exchange can be performed. Each option will have a particular, predefined expiration. For example, certain index-based options expire in the morning of the third Friday of the month. Most stock-based options expire in practice at the end of the third Friday of the month (although they will list the Saturday as the formal expiration day). More recently, several weekly-based option contracts have been made available for some of the most traded stocks and indices. Each options contract makes it clear when expiration is due, and investors need to plan accordingly on what to do before the expiration date.

CHAPTER 1 ▪ OPTIONS CONCEPTS

*Settlement*: The settlement is the agreed-on result of the option transaction at the specific time when the option contract expires. The particular details of the settlement depend on the type of underlying asset. For example, options on common stock settle at expiration day, when the owner of the option needs to decide to sell (for puts) or buy (for calls) a certain quantity of stock. For index-based options, the settlement is normally performed directly in cash, determined as the cash equivalent for a certain number of units of the index. Some options on futures may require the settlement on the underlying commodity, such as grain, oil, or sugar. Investors need to be aware of the requirement settlement for different option contracts. Trading brokerages will typically let investors know about the steps required to settle the options they're currently holding.

*Selling Options:* An investor can buy or sell a call option. When doing so, it is important to understand the difference between these two situations. For option buyers, the goal is to profit from the possible increase (in the case of calls) or the decrease (in the case of puts) in value for the underlying. For option sellers, on the other hand, the goal is to profit from the lack of movement (increase for calls or decrease for puts). So, for example, if you sell calls against a stock, the intent is to profit in the case that the stock decreases in price or stays at the same price until expiration. If you sell a put option, the goal is to profit when the stock increases in price or stays at the same price until expiration.

*Option exercise*: An option contract can be used to buy or sell the underlying asset as dictated by the contract specification. This process of using the option to trade the underlying asset is called *option exercising*. If the option is a call, you can exercise it and buy the underlying asset at the specified price. If the option is a put, you can use the option to sell the underlying asset at the previously specified price. The price at which the option is exercised is defined by the contract. For example, a call option for AAPL stock with a $100 strike allows its owner to buy the stock for the strike price, independent of the current price of AAPL.

*Exercise style*: Options contracts can have different exercise styles based on when exercising is allowed. There are two main types:

- *American options*: Can be exercised at any time until expiration. That is, the owner of the option can decide to exercise it at any moment, as long as the option has not expired.

- *European options*: Can be exercised only upon expiration date. This style is more common for contracts that are settled directly on cash, such as index-based options.

An option is defined as a derivative of an underlying instrument. The *underlying instrument* is the asset whose price is used as the basic value for an option contract. There is no fixed restriction on the type of asset used and the underlying asset for an option contract, but in practice options tend to be defined based on openly traded securities. Examples of securities that can be used as the underlying asset for commonly traded option contracts include the following:

- *Common stock*: Probably the most common way to use options is to trade call or put options on common stock. In this way, you can profit largely from small price changes in stocks of public companies.

- *Indices*: An index, such as the Dow Industrials or the NASDAQ 100, can be used as the underlying for an options contract. Options based on indices are traditionally settled on cash (as explained below), and each unit of value corresponds to multiples of the current index value.

- *Currencies*: A currency, usually traded using Forex platforms, can also be used as the underlying for option contracts. Common currencies such as the U.S. Dollar, Euro, Japanese Yen, and Swiss Franc are traded 24 hours a day. The related options are traded on lots of currencies, which are defined according to the relative prices of the target currencies. Expiration varies similarly to stock options.

CHAPTER 1 ■ OPTIONS CONCEPTS

- *Commodities*: Options can also be written on commodities contracts. A commodity is a common product that can be traded in large quantities, including agricultural products such as corn, coffee, and sugar; fuels such as gasoline and crude oil; and even index-based underlying assets such as the S&P 500. Options can be used to trade such commodities and trading exchanges now offer options for many of the commodity types.

- *Futures*: These are contracts for the future delivery of a particular asset. Many of the commodity types discussed previously are traded using future contracts, including gasoline, crude oil, sugar, coffee, and other products. The structure of future contracts is defined to simplify the trade of products that will only be available in a due period, such as next fall, for example.

- ETFs *(Exchange Traded Funds) and ETN (Exchange Traded Notes):* More recently, an increasing number of funds have started to trade using the same rules applicable to common stocks in standard exchanges. Such funds are responsible for maintaining a basket of assets, and their shares are traded daily on exchanges. Examples of well-known ETFs include funds that hold components of the S&P 500, sectors of the economy, and even commodities and currency.

Options trading has traditionally been done on stock exchanges, just like other forms of stock and future trading. One of the most prominent options exchange is the Chicago Board Options Exchange. Many other exchanges provide support and liquidity for the trading of options for many of the instruments listed here.

The techniques described in this book are useful for options with any of these underlying instruments. Therefore, you don't need to worry if the algorithms are applied to stock options of the futures options, as long as you consider the possible peculiarities of these different contracts, such as their expiration and exercise.

Options can also be classified according to the relation between the strike price and the price of the underlying asset. There are three cases that are typically considered:

- An option is said to be *out of the money* (OTM) when the strike price is above the current price of the underlying asset (for call options) or when the strike price is below the current price of the underlying asset (for put options).

- An option is said to be at the money (ATM) when the strike price is close to the current price of the underlying asset.

- An option is said to be in the money (ITM) when the strike price is below the current price of the underlying asset (for call options) or when the strike price is above the current price of the underlying asset (for put options).

Notice that OTM options are cheaper than a similar ATM option, since the OTM options (being away from the current price of the underlying) have a lower chance of profit than ATM options. Similarly, ATM options are cheaper than ITM options, because ATM options have less probability of making money than other ITM options. Notice that, when considering the relation between strike price and underlying price, the option price reflects the probability that the option will generate any profit.

A related concept is the *intrinsic value* of an option. The intrinsic value is the part of the value of an option that can be derived from the difference between strike price and the price of the underlying asset. For example, consider an ITM call option for a particular stock with a strike of $100. Assume that the current price for that stock is $102. Therefore, the price of the option must include the $2 difference between the strike and the price of the underlying, since the holder of a call option can exercise it and have an immediate profit of $2. Similarly, ITM put options have intrinsic value when the current price of the underlying is above the strike price, using the same reasoning.

The *break-even price* is the price of the underlying at which the owner of an option will start to make a profit. The break-even price has to include not only the potential profit derived from the intrinsic value, but also the cost paid for the option. Therefore, for an investor to make a profit on a call option, the price of the underlying asset has to rise above the strike plus any cost paid for the option (and similarly it has to drop below the strike minus the option cost for put options). For example, if an $100 MSFT call option has a cost of $1, then the investor will have a net profit only when the price of MSFT rises above $101 (and this without considering transaction costs).

As part of the larger picture of investing, options have assumed an important role due to their flexibility and their profit potential. As a result, new programming problems introduced by the use of options have come to the forefront of the investment industry, including banks, hedge funds, and other financial institutions. As you will see in the next section, C++ is the ideal language to create efficient and elegant solutions to the programming problems occurring with options investing.

## Option Greeks

One of the characteristics of derivatives is the determination of quantitative measures that can be essential in the analysis and pricing of the derivative product. In the case of options, the quantitative measures are called *Greeks*, because most of these measures are named after Greek letters. Many of these Greek quantities correspond to the variation of the price with respect to one or more variables, such as time, strike, or underlying price.

The most well known option Greek is *delta*. The delta of an option is defined as the amount of change in the price of an option when the underlying changes by one unit. Therefore, delta represents a rate of change of the option in relation to the change in the underlying, and it is essential to understand price variation in options. Consider, for example, an option for IBM stock that expires in 30 days. The strike price is $100, and the stock is currently trading at $100. Suppose that the price of the stock increases by $1. It is interesting to calculate the expected change in the option price. It turns out that when the underlying price is close to the strike price, the delta of an option is close to 0.5. Expressing this in terms of probabilities, it means the value of the stock is equality probable to go up or down. Therefore, it makes sense that the change per unit of price will be just half of the change in the underlying asset.

The value of delta increases as the option becomes more and more in the money. In that case, the delta gets close to one, since each dollar of change will have a larger impact in the intrinsic value of the option. Conversely, the value of delta decreases as the option becomes more and more out of the money. In that case, delta gets closer to zero, since each dollar of change will have less impact on the value of an option that is out of the money.

The second option Greek that is related to delta is *gamma*. The gamma of an option is described as the rate of change of delta with a unit change in price of the underlying. As you have seen, delta changes in different ways when the option is in the money, out of the money, or at the money. But the rate of change of delta will also vary depending on other factors. For example, delta will change more quickly if the option is close to expiration, because there is so little time for a movement to happen. To see why this happen, consider an option delta 30 days before expiration and one that is one day before expiration. Delta is also dependent on time, because an option close to expiration has less probability of change. As a result, the delta will move from zero to one slowly if there are 30 days to go, because there is still plenty of time left for other changes. But an option with only one day left will quickly move from close to zero delta to near one, since there is no time left for future changes. This is described by saying that the first option has lower gamma than the second option. Other factors such as volatility can also change an option gamma. Figure 1-2 illustrates the value of gamma for a particular option at different times before expiration.

# CHAPTER 1 ■ OPTIONS CONCEPTS

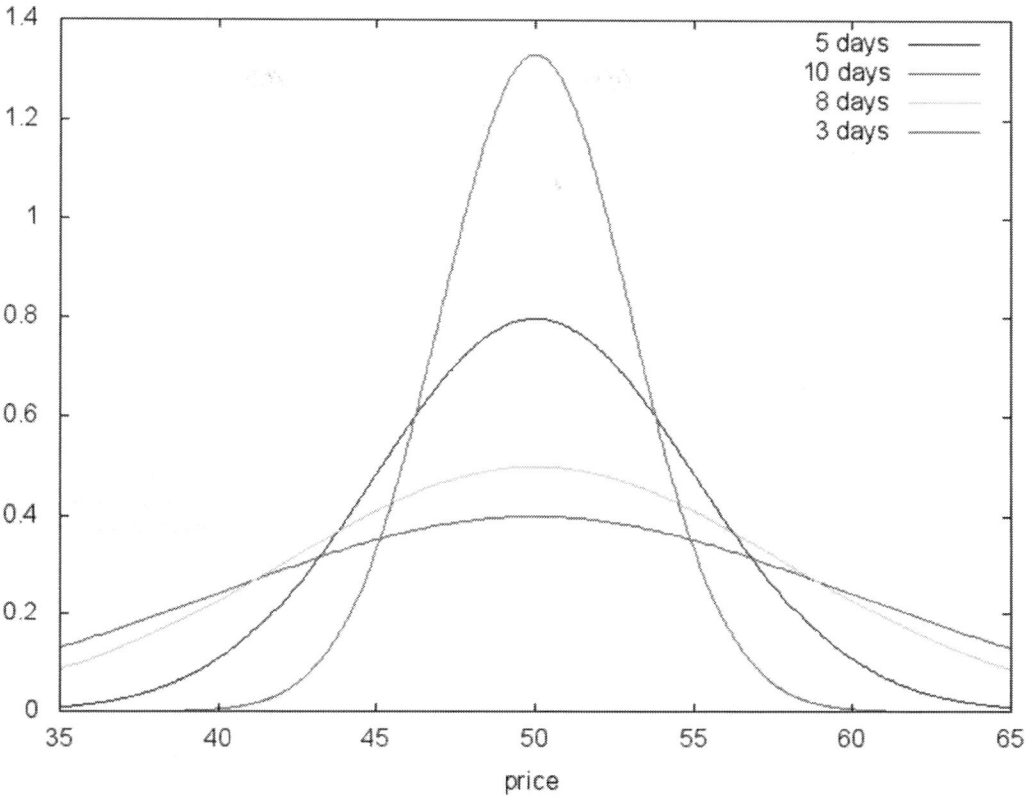

*Figure 1-2. Value of gamma at different dates before expiration*

Another option Greek that is closely related to time is *theta*. The theta of an option is proportional to the time left to expiration, and its value decays when it gets closer to the expiration date. You can think of theta as a measure of time potential for the option. For option buyers, higher theta is a desirable feature, since buyers want more probability of changes for their options. On the other hand, option sellers benefit from decreased theta, so short-term options are ideal for sellers due to the lower theta.

Finally, we have an option Greek that is not really named after a Greek letter: *vega*. The vega of an option measures the amount of volatility of the underlying asset that is priced into an option. The higher the volatility, the more expensive an option has to be in order to account for the increased possibility of pricing changes. The differential equations that define the price of an option (as you will see in future chapters) take into account this volatility. Vega can be used to determine how much relative volatility is embedded in the option price, and an important use of this measure is to help option buyers and sellers determine if this implied volatility is consistent with their expectations for the future of the underlying prices.

There are other option Greeks that have been used in the academic community and in some financial application; however, they are not as widespread as the ones mentioned here. You can see a summary of the most commonly used option Greeks in Table 1-2.

*Table 1-2. Option Greeks and Their Common Meanings*

| Greek | Meaning |
| --- | --- |
| Delta | The rate of change of the option value with respect to the price of the underlying asset. |
| Gamma | The rate of change of delta with respect to the price of the underlying asset. |
| Rho | The change of the price of the option with respect to changes in interest rates. |
| Theta | The rate of change in the option value with respect to time left to expiration. |
| Vega | The rate of change of the option value with respect to the volatility of the underlying asset. |
| Lambda | The rate of change in the option value with respect to percent changes in the price of the underlying asset. |

# Using C++ for Options Programming

C++ has unique features that make it especially useful for programming software for the financial industry. Over the years, developers have migrated to C++ as a practical way to meet the requirements of numeric, real-time algorithms used by the investment community. When it comes to creating decision support software for fast-paced investment strategies, it is very difficult to beat the C++ programming language in the areas of performance and flexibility.

While it is true that several programming languages are available for the implementation of financial software, very few of them provide the combination of advantages available when using C++. Let's look at some of the areas where C++ provides a unique advantage when compared to other programming languages that could be used to implement investment software.

## Availability

When looking for a programming language to implement investment software, one of the first concerns you need to address is the ability to run the code in a variety of computational environments. Targets for such investment software can range from small and mobile processors to large-scale parallel systems and supercomputers. Moreover, it is not uncommon to have to interact with different operating systems, including the common software platforms based on Linux, Windows, and MacOS X.

Because modern computer systems are so heterogeneous, it makes economic sense to use languages that can be employed in a large variety of hardware and software configurations with little or no source code modifications. Financial programmers also work on different platforms, which makes it even more interesting to use software that can run in different computers and operating systems with little or no changes.

A strong characteristic of C++ is its wide availability over different platforms. Due to its early success as a multi-paradigm language, C++ has been ported to nearly any imaginable software and hardware combination. While other mainstream languages such as Java require the implementation of a complex runtime environment for proper operation, C++ was designed from the beginning with simplicity and portability in mind. The language does not require a runtime system, and only a minimal support system, provided by the C++ standard library, needs to work in order to support a new target. Therefore, it is relatively easy to port C++ compilers and build systems to new platforms with minimal changes.

Another advantage is the availability of multiple compilers provided by commercial vendors as well as free software. Given the importance of C++ software, it is possible to find compilers with both free and under commercial licenses, so that you can use the scheme that best suits your objectives. Open source developers can use state-of-the-art free compilers such as gcc and LLVM cc. Commercial groups, on the other hand, can take advantage of compilers licensed by companies such as Intel and IBM.

## Performance

It is fact that programmers using C++ benefit from the high performance provided by the language. Because C++ was explicitly designed to require a minimum amount of overhead in most platforms, typical C++ programs run very efficiently, even without further optimization steps. Programs naturally coded in C++ will frequently outperform code created in other languages, even when this software has been heavily optimized.

Part of the performance advantage provided by C++ is a result of mature compilers and other building tools. Since C++ is such a well-established language, major companies as well as well-known open source projects have created optimized compilers for the language. Common examples include gcc, Visual C++, LLVM cc, and Intel cc. Such compilers provide huge speed improvements in typical running time, frequently beating non-optimized (and even optimized) code that is produced by other languages.

When considering performance, C++ shares the same philosophy of the C programming language. The general idea is to provide high-level features, while avoiding whenever possible any overhead on the implementation of such features on common processors. This means that the features provided by C++ generally match very closely with low-level processor instructions.

Other solutions for improved performance in C++ include the use of templates in addition to runtime polymorphism. With templates, the compiler can generate code that matches the types used in a particular algorithm exactly. In this way, programs can avoid the large overhead of polymorphic code, which need to made different runtime decisions depending on the particular type. Programmers can control algorithms in a much finer grained scale when using templates, while still retaining the ability to use high-level types.

Last but not least, C++ simplifies the use of memory and other resources with the help of smart pointers and other techniques based on RAII (Resource Acquisition Is Initialization). These techniques allow C++ programs to control memory usage without having to rely on a runtime GC (garbage collection) system. By using such strategies, C++ programmers can considerably reduce the overhead of frequently used dynamic algorithms, without the need to resort to manual bookkeeping of memory and other resources.

## Standardization

Another great advantage of C++ is that it's based on an international standard, which is recognized by practically every software vendor. Unlike some languages that are practically defined by an actual implementation or controlled by a powerful company, C++ has for decades being defined as the result of the C++ committee, with representatives from major companies and organizations that have an interest in the future development of the language.

In fact, some of the big financial companies also have representatives in the C++ committee. This means that the future of C++ is not controlled by a single institution, such as what happens with Java (controlled by Oracle), C# (controlled by Microsoft), or Objective-C and Swift (controlled by Apple). The fact that the standards committee has members from several organizations protects its users from commercial manipulation that would benefit a single company, to the detriment of the larger community of users.

The C++ standards committee has been effective in improving the language in ways that address many of the modern needs of programmers. For example, the last two version of the language standard (C++11 and C++14) introduced many changes that simplify common aspects of programming, such as simpler initialization methods, more advanced type detection, and generalized control structures.

The standard library has also been the target of many improvements over the last few years. A main focus has been the introduction of containers and smart pointers, which can be used to simplify a large part of modern applications. The standard library also has been augmented to support parallel and multithreaded algorithms, using primitives that can be reused on different operating systems and architectures.

It is necessary to remember that the standardization process has a few drawbacks too. One of the issues is the time it takes to introduce new features. Since the standardization process requires a lot of organization and formal meetings, it takes several years before a new version of the standard is approved. Also, there is the risk of including features that go against the previous design of the language. In this case, however, the committee has been very careful in introducing only features that have been thoroughly tested and considered to improve the language according to its philosophy.

In general, having a standardized language has certainly helped the C++ community to grow and improve the whole programming ecosystem over the last few decades. This is just another reason why financial developers have embraced C++ as a language suitable for the implementation of options and derivative-based financial algorithms.

## Expressiveness

Last, but not least, C++ is multi-paradigm language that provides the expressiveness and features necessary for the implementation of complex software projects. Unlike some languages, which define themselves as following a single programming paradigm (such as object-oriented or functional), C++ allows use of multiple paradigms in a single application. In this way, you can use the best approach for problem solving, independent of the necessary implementation techniques: object-oriented programming, functional programming, template-based programming, or just simple structured programming.

Because C++ allows programmers to express themselves using different paradigms, it makes easier to find a solution that matches the problem at hand, instead of requiring changes to the way you think in order to match the requirements of the language. For example, a language such as Java, which is designed as object-oriented, requires programmers to create code based on objects and classes even when this does not match the underlying requirements of the problem. In C++, on the other hand, you have a choice of using OO techniques as well as functional or even more traditional structured techniques, if this is what your algorithm requires.

The fact that you can use different techniques for different parts of your application also improves your ability to concentrate on algorithms, instead of on programing techniques. Sometimes using a template-based strategy is the easiest way to achieve a particular algorithmic goal, and C++ allows you to do that without getting in your way. Other parts of the application may benefit from using objects, such as the GUI code. In each case, it is important to be able to express algorithms in the most natural way.

In this book, you will have the opportunity to use many of the features of C++ in different contexts. It will be clear that some features such as object-oriented programming are best used with a particular class of problems, whereas functional techniques may be the best approach in other situations. The fact that the C++ language provides the flexibility to tackle such problems is a distinct advantage.

# Modeling Options in C++

In this section, you will learn how to code a basic class that can later be used as a starting point for options analysis and trading. In this first example, you will see a C++ class that can be used as the basis for a framework for options value calculation. The class is name `GenericOption`, since it can be used for any type of underlying, and for calls and puts. Before I present how the class works, let's review a basic concept of class design that is unique to the C++ language, and which will be followed in the examples of this book.

## Creating Well-Behaving Classes

One of the most important parts of designing classes for C++ is to make sure that they can combine appropriately with other libraries in the system. In particular, the C++ standard library, which includes the STL (standard template library), is the most important set of classes that you will encounter when developing C++ applications. It is essential that your classes play well with the classes and templates provided by the standard library.

CHAPTER 1 ▓ OPTIONS CONCEPTS

To work properly with other parts of the C++ library, classes need to define (or use the default definition for) the four special member functions. These member functions are mainly used to create and copy objects, and are required in general to guarantee their proper behavior. These four special member functions are:

- *The constructor*: Each class can have one or more constructors that define how to initialize objects of the class. The constructor is named after the class, and it can be overloaded so that you can create classes with different parameters. The constructor that receives no parameters is also known as the default constructor, and the compiler automatically creates it if you don't supply one. Most of the time you should avoid using the default constructor because it doesn't properly initialize the native C++ types, such as the double and int variables. To avoid such issues, you should always provide a constructor for new classes.

- *The copy constructor*: This specialized constructor performs a initialization similar function; however, it is called only when creating objects from an existing object. The C++ compiler can also generate a default copy constructor, which copies the values stored in the original object into the new object. However, the default copy constructor also has a problem: it doesn't know the semantics of some values stored in the object. This causes problems when you're storing a pointer to allocated memory or some object that shouldn't be copied. To avoid such problems, you should provide your own definition for the copy constructor. The best way to avoid such issues is to always write a copy constructor for new classes.

- *The destructor*: A destructor defines how data used by the class is released when the object is destroyed. Like the other special member functions, the compiler generates a default empty constructor. You should add your own constructor to properly handle the release of private data.

- *The assignment operator*: When copying data between objects, the assignment operator is invoked automatically. Even though this special method is not equivalent to a constructor, it does similar work. Therefore, you should apply the same strategy when dealing with the assignment operator and make sure that it properly handles initialization and copies the required data members.

To avoid potential problems with C++ classes, it is best to include these four member functions in all the classes you create. They are pretty straightforward. The only member function that needs further explanation is the assignment operator. Suppose that you're implementing a class called CppClass. The assignment operator would read as follows:

```
GenericOption &GenericOption::operator=(const GenericOption &p)
{
    if (this != &p)
    {
        m_type = p.m_type;
        m_strike = p.m_strike;
    }
    return *this;
}
```

The reason for this check is that you don't want to perform the private data member assignment unless the objects in the left and right side of the assignment operator are different:

```
if (this != &p)
```

While performing the auto-assignment might not be a problem for some types of variables (especially for native types), it can be time consuming for complex objects that need to perform several steps during initialization and release. For example, if a member is a large matrix, the assignment may trigger an expensive copy operation that is unnecessary.

## Computing Option Value at Expiration

The example class `GenericOption` provides only the minimum necessary to calculate the value of options at expiration. The first thing you should notice about this class is that it follows the recommended practice described in the previous section. Therefore it contains a constructor, a copy regular constructor, a destructor, and an assignment operator.

The main constructor of `GenericOption` does very little and is responsible only for the initialization of private variables. Although this is normal in a simple class, using constructors with very few responsibilities is a pattern that should be adopted in most cases. Since constructors are used in many places in the C++ language, it is important to make them as fast as possible and relegate any complex operations to member functions that can be called after the object is created.

■ **Tip**   Avoid complex constructors in the classes you design. Constructors are frequently called for the creation of temporary objects and when passing parameters by value, for example. Complex constructors can make your code run slower and make classes harder to maintain.

There are two types of options recognized by the `GenericOption` class. This is defined by the enumeration `OptionType`, which contains the values `OptionType_Call` and `OptionType_Put`. Depending on the value passed to the constructor, the object will behave accordingly as a call or as a put option. The constructor also requires the strike value of the option and the cost of the option when it was bought. You will see later in this book how this option cost can be calculated from other parameters, but for now you can assume that the cost of the option is provided by the exchange.

The main functionality of the class is contained in two member functions: `valueAtExpiration` and `profitAtExpiration`. The first member function simply calculates the value of the option at the time of expiration, which in this case is the same as the intrinsic value. To perform this calculation, it needs to know the current price of the underlying asset. The member function `valueAtExpiraton` first needs to determine if the option is a put or a call. In the case of a put, it takes the difference between the current price and the strike price as the value, with the value being zero when the strike is lower than the current price. In the case of a call, this calculation is reversed, with the value being zero when the strike price is higher than the current price. The full calculation can be coded as follows:

```
double GenericOption::valueAtExpiration(double underlyingAtExpiration)
{
    double value = 0.0;

    if (m_type == OptionType_Call)
    {
```

```
        if (m_strike < underlyingAtExpiration)
        {
            value = underlyingAtExpiration - m_strike;
        }
    }
    else   // it is an OptionType_Put
    {
        if (m_strike > underlyingAtExpiration)
        {
            value = m_strike - underlyingAtExpiration;
        }
    }
    return value;
}
```

The profitAtExpiration function is similar to valueAtExpiration, but it also considers the price that was paid by the option. Thus, a profit in the option is achieved only after it surpasses the break-even price (for call options). The calculation uses the m_cost member variable to determine the price paid by the option, and it returns the net profit of the option (without considering transaction costs). The function can be coded as follows:

```
double GenericOption::profitAtExpiration(double underlyingAtExpiration)
{
    double value = 0.0;
    double finalValue = valueAtExpiration(underlyingAtExpiration);
    if (finalValue > m_cost)
    {
        value = finalValue - m_cost;
    }
    return value;
}
```

## Complete Listing

The complete code for the example described previously is shown in Listing 1-1. The code is split into a header file called GenericOption.h and an implementation file called GenericOption.cpp.

***Listing 1-1.*** Interface of the GenericOption class

```
//
//  GenericOption.h

#ifndef __CppOptions__GenericOption__
#define __CppOptions__GenericOption__

#include <stdio.h>
```

CHAPTER 1 ■ OPTIONS CONCEPTS

```cpp
//
// option types based on direction: call or put
enum OptionType {
    OptionType_Call,
    OptionType_Put
};

//
// class the represents a generic option
//
class GenericOption {
public:
    GenericOption(double strike, OptionType type, double cost);
    GenericOption(const GenericOption &p);
    ~GenericOption();
    GenericOption &operator=(const GenericOption &p);

    double valueAtExpiration(double underlyingAtExpiration);
    double profitAtExpiration(double underlyingAtExpiration);
private:
    double m_strike;
    OptionType m_type;
    double m_cost;
};

#endif /* defined(__CppOptions__GenericOption__) */
```

***Listing 1-2.*** Implementation of the GenericOption class

```cpp
//
//  GenericOption.cpp

#include "GenericOption.h"

using std::cout;
using std::endl;

GenericOption::GenericOption(double strike, OptionType type, double cost)
: m_strike(strike),
  m_type(type),
  m_cost(cost)
{

}

GenericOption::GenericOption(const GenericOption &p)
: m_strike(p.m_strike),
  m_type(p.m_type),
  m_cost(p.m_cost)
{

}
```

14

```cpp
GenericOption::~GenericOption()
{
}

//
// assignment operator
GenericOption &GenericOption::operator=(const GenericOption &p)
{
    if (this != &p)
    {
        m_type = p.m_type;
        m_strike = p.m_strike;
        m_cost = p.m_cost;
    }
    return *this;
}

//
// Computes the value of the option at expiration date.
// Value depends on the type of option (CALL or PUT) and strike.
//
double GenericOption::valueAtExpiration(double underlyingAtExpiration)
{
    double value = 0.0;

    if (m_type == OptionType_Call)
    {
        if (m_strike < underlyingAtExpiration)
        {
            value = underlyingAtExpiration - m_strike;
        }
    }
    else  // it is an OptionType_Put
    {
        if (m_strike > underlyingAtExpiration)
        {
            value = m_strike - underlyingAtExpiration;
        }
    }
    return value;
}

//
// return the profit (value at expiration minus option cost)
//
double GenericOption::profitAtExpiration(double underlyingAtExpiration)
{
    double value = 0.0;
    double finalValue = valueAtExpiration(underlyingAtExpiration);
```

```cpp
        if (finalValue > m_cost)
        {
            value = finalValue - m_cost;
        }
        return value;
    }

    int main()
    {
        GenericOption option(100.0, OptionType_Put, 1.1);
        double price1 = 120.0;
        double value = option.valueAtExpiration(price1);
        cout << " For 100PUT, value at expiration for price "
             << price1
             << " is "
             << value << endl;
        double price2 = 85.0;
        value = option.valueAtExpiration(85.0);
        cout << " For 100PUT, value at expiration for price "
             << price2
             << " is "
             << value << endl;

        // test profitAtExpiration
        auto limit = 120.0;
        for (auto price = 80.0; price <= limit; price += 0.1)
        {
            value = option.profitAtExpiration(price);
            cout << price << ", " << value << endl;
        }

        return 0;
    }
```

## Building and Testing

To compile the code presented in the last section, you need a standards compliant C++ compiler. I have tested this code with gcc and LLVM cc, although most compilers should work without any problems. Here are the commands that I used to compile this on my machine:

```
gcc -o GenericOption.o -c GenericOption.cpp

gcc -o GenericOption GenericOption.o
```

The executable file can then be used to run the sample application like this (I used the bash shell to run the application on UNIX):

```
$ ./GenericOption

For 100PUT, value at expiration for price 120 is 0
For 100PUT, value at expiration for price 85 is 15

80, 20
80.1, 19.9
80.2, 19.8
80.3, 19.7
80.4, 19.6
80.5, 19.5
80.6, 19.4
80.7, 19.3
...
```

You can check the output to determine if the results match your expectations. I used the data to create a chart with the results, as shown in Figure 1-3. Since the example is a put, notice that the profit is negative for any price higher than the break-even price of $98.90. Below that value, the profit rises steadily, attaining its maximum value at price $0 (not shown in the chart).

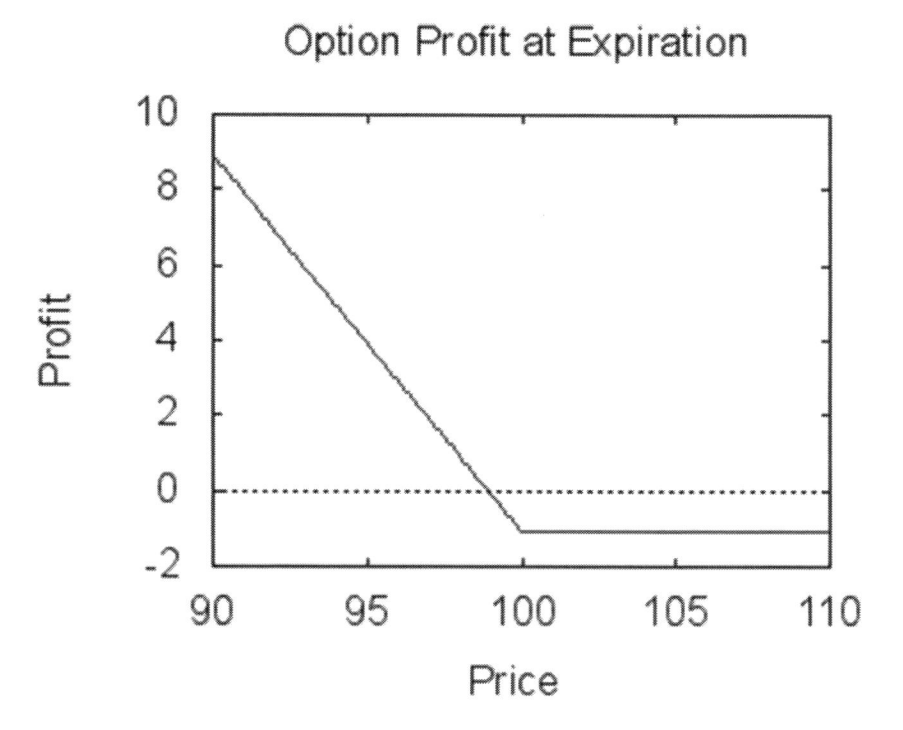

*Figure 1-3.* *Profit chart calculated with the GenericOption class for sample option with strike price $100*

## Further References

In this chapter, I provided an introduction to most common concepts of options investing, and how C++ programmers can model them. You can turn to several other sources for further clarity on the concepts introduced in this chapter. If you need additional information on options and related financial investments, here are a few books that cover not only the basics but also the mathematical details of options investing:

- *Option Volatility & Pricing,* by Sheldon Natenberg, McGraw Hill, 1994. This is the standard reference on options and their properties. This book explains in great detail how options are defined, how option Greeks work, and their basic economic interpretation.
- *Investment Science,* by David Luenberger, Oxford University Press, 1998. This is an undergraduate-level book that describes the basic theory of investment. Most of the book explains the fundamentals of fixed income investments, but the included algorithms can be used for other common problems in finance.
- *Mathematics for Finance,* by Marek Capinski and Tomasz Zastawniak, Springer Press, 2011. This book is more for the mathematically inclined. It explains not only the basics of fixed income investments, but also gives a lot of mathematical methods that are useful in their analysis. Many of these techniques are also used in analysis of options-based investments.
- *Investments* by Zvi Bodie, Alex Kane, and Alan J. Marcus, McGraw Hill/Irwin, 2004. This is a standard textbook on investment theory that explains, among other topics, the ideas behind equity-based investments and their derivatives.

## Conclusion

In this chapter, I provided an overview of the themes and ideas that will be discussed in the remainder of the book. Options are basic financial vehicles that can serve multiple investment goals such as providing risk protection, supplying short-term income, or serving as a speculation method based on perceived future prices of a financial instrument.

I started with a basic description of options and how they fit in the landscape of the investment industry. You learned the most important properties of options and how they define standard contracts that are traded by stock, futures, and commodity exchanges. I also described how this information may be useful to software engineers who want to create solutions for the financial industry using C++ as the main implementation language.

You have seen how options can be described by options Greeks, a set of standard attributes associated with option contracts that can be used to determine several properties of the option. In particular, these option Greeks are useful for evaluating the price at which options should be bought and sold, as you will see in the algorithms introduced in the later part of this book.

This chapter also discussed the advantages of C++ as a language for financial and options-related programming. Many of the features of C++ make it an ideal language to implement algorithms and large-scale software packages to analyze and trade options. You have seen an example C++ class that can be used to compute the profit or loss for a single option contract.

In the next chapter, you will learn about derivatives in general and how they expand on the ideas of common options. You will also see how such financial derivatives can be modeled using the C++ language.

# CHAPTER 2

# Financial Derivatives

*Derivative* is a general term for contracts that have their price based on the properties of an underlying asset. In particular, options are a standardized type of derivatives that give the right to buy or sell the underlying asset at a particular price. Unlike options, however, general derivatives include a large number of non-standard features that allow them to be created even for illiquid assets such as corporate credit risk or real estate mortgages.

In the last decades, the financial industry has created and popularized a multitude of derivatives to collateralize disparate assets, including items such as fixed income instruments, mortgage-backed securities, and risk of default. Pricing and trading these derivatives is a large part of the work performed daily in the trading desks of large banks and by quantitative programmers throughout the world.

This chapter focuses on characteristics of general derivatives and presents a few C++ techniques that are useful to model specific aspects of these financial instruments. The topics in this chapter also introduce you to what you will learn in more depth in the remainder of this book. The main topics covered in the chapter are:

- *Models for derivative pricing:* You will learn the basic ideas used to determine the price of various derivatives along with a few examples of how they work.
- *Credit default swaps:* A particular type of derivative where investors want to buy protection against the default of a third-party institution.
- *Interest rate derivatives:* A derivative in which the underlying asset is an interest rate that is paid in predefined time periods.
- *FX derivatives:* A quick introduction to some exotic derivative contracts.
- *A Monte Carlo model for derivatives:* You will explore a simple computation of Monte Carlo models for pricing derivatives.
- *Using the STL for derivative pricing:* Using the STL makes it possible to create fast containers for generic objects, without incurring runtime inefficiencies.

## Models for Derivative Pricing

In the last chapter you learned some basic information about options and how to use C++ to work with these simple contracts. Recall that an *option* is a kind of financial derivative that is traded on exchanges and uses a standard agreement between buyers and sellers. General derivatives, however, are not restricted to the fixed requirements of a simple option contract. In this chapter, you will learn about more complex derivatives, including how they are handled in the financial community.

## CHAPTER 2 ■ FINANCIAL DERIVATIVES

In its general sense, a financial derivative is just a contract that assigns a value to a particular set of rights linked to an underlying asset. For example, options give the right to buy or sell an asset such as a stock or a commodity. But more complex derivatives can be created if you just a more exotic transaction between buyers and sellers. For example, credit default swaps are contractual exchanges that require a payment to occur only when a particular entity is in default. For another example, collateralized debt obligations will require payments that depend on the risk level of certain borrowers.

The common aspect shared between different derivatives is the way their prices are modeled, that is, the mathematical characteristics of price changes for these instruments. All derivatives that are traded in the market can be analyzed using a generalized random walk model that was discovered and applied in the 20th century by American economists. Such a model for derivative pricing and their associated mathematical equations were developed and popularized by Robert Merton, in a work that was itself a generalization of the Nobel prize winning Black-Scholes model for options pricing.

In a random walk model, the prices of securities are studied under the assumption that the changes are random. That is, prices can move up or down by a random value that is given by a normal distribution, as shown in Figure 2-1. While this is only an approximation of the complex market behavior, it is so close to what has been observed in the marketplace that models based on random walks have been extremely successful. These models have been frequently used in the financial industry to accurately determine prices for options and more complex derivatives. As a result, most of what you will learn in this book is in some way or another related to how to explore this pricing model when creating software to analyze and trade derivatives.

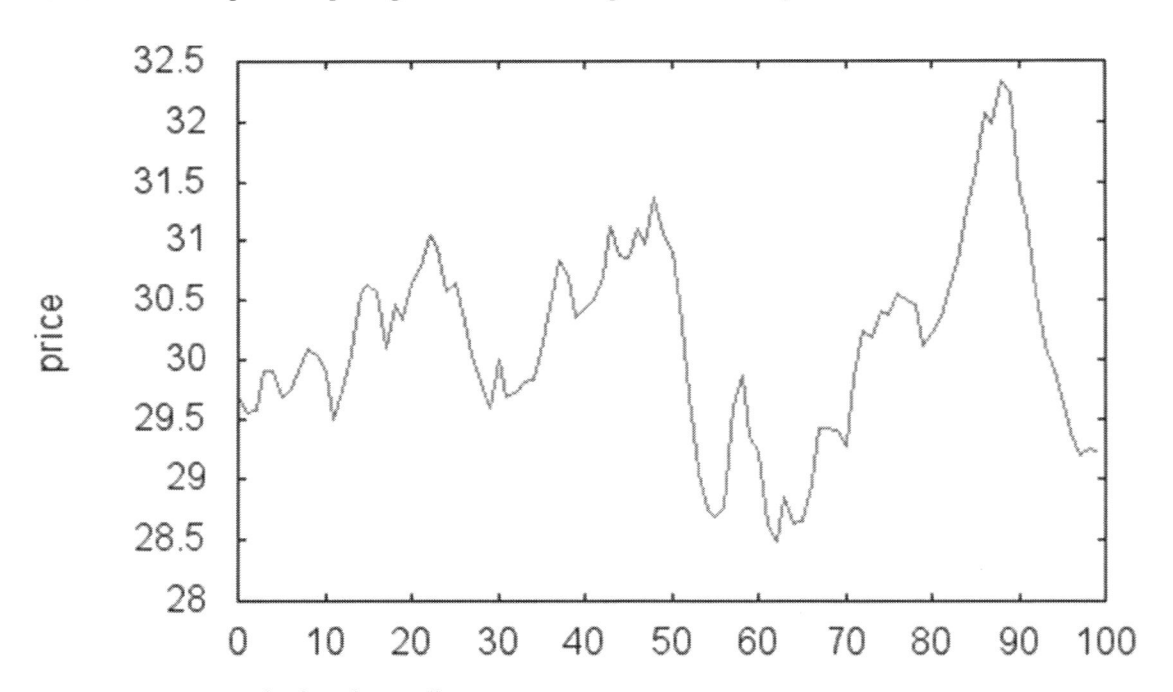

*Figure 2-1. An example of random walk*

The first thing to understand about the random walk model for derivative pricing is that it results in a set of equations that determine the behavior of prices as time passes. This equation is, by the nature of its assumptions probabilistic, but it can be solved to give a value for the fair price of a particular investment instrument.

The *fair price*, according to economic assumptions, is the price at which neither the sellers nor the buyers would have an unfair advantage. In other words, both sides in the transaction are satisfied with the result, and there is no known way to extract more value from one of the sides in the transaction without breaking this equilibrium. Because the model used is probabilistic, this also means that each side of the transaction has the same probability of making money after the transaction is concluded. This fair price element of the model allows you to calculate a fixed value using only a probabilistic assumption about future expectations.

*Table 2-1. A List of Common Derivatives*

| Derivative Type | Description |
| --- | --- |
| Credit default swaps | A contract that pays its holder in the case of a default of a target corporation. |
| Collateralized debt obligations | A financial product where debt is paid to investors according to levels of risk collateral from borrowers. |
| FX derivatives | A derivative where the underlying asset is composed of foreign currencies, with prices varying according foreign exchange trading. |
| Interest rate Derivatives | A derivative in which the underlying asset is an interest rate that is paid in predefined time periods. |
| Mortgage-backed security | A type of derivative that is defined in terms of mortgage contracts. |
| Energy derivative | Derivative in which the underlying asset is an energy product or asset, such as oil, natural gas, coal, or electricity. |
| Inflation derivative | Derivative contracts that have a price defined by the level of inflation in a particular economy. |

Another consequence of fair prices is that the resulting model allows no *arbitrage*. Arbitrage is a method of making money in financial markets where you buy some asset for a price and immediately sell it for a higher price for a sure profit. This kind of arbitrage cannot be allowed in a financial model, because it indicates that the original price was unfair for at least one of the participants. It also corresponds to the known fact that, in liquid and free markets, opportunities for arbitrage will be non-existent or disappear as soon as they are identified.

## Credit Default Swaps

A credit default swap is a derivative that allows investors to bet on the solvency of a particular institution. In this case, the underlying asset is defined as the value of a business minus the liabilities it currently has. *Solvency* is then defined as the situation in which the value of the business is superior to its liabilities.

Credit default swaps have been used as a way to protect large corporations against the risk of default of a counterpart, which is a common risk suffered by contracts with large institutions. For example, the 2008 financial meltdown proved that counterpart risk is very difficult to avoid when only a few participants dominate a large part of the market. The ability to use mathematical techniques to model this type of risk is therefore very important for institutions that deal with such large-scale operations.

In the recent years, most banks and other investment institutions have become active in the development of CDS models as a way to mitigate such risks. Much of the software for solving CDS pricing models is based on modern C++, which you will learn in the next chapters.

## Collateralized Debt Obligations

A *collateralized debt obligation* is a financial derivative product based on the cash flow of a collection of loans. The collateralization process makes it possible to split the cash flows among different investors based on the characteristics of individual loan originations.

In particular, CDOs are used to split cash flows based on risk of loans. Parts of the cash flow are classified as low risk (for example, loans that are labeled as AAA by credit rating institutions) and sold for higher profit, while other parts of the package are sold as higher-risk investments. CDOs have acquired a bad public reputation during the financial crisis, but they remain a valuable tool for defining the risk associated with particular investment classes.

CDO pricing relies heavily on the derivative pricing techniques that will be discussed in this book. The development of Black-Scholes-Merton methods gave institutions the ability to price more complicated products using similar ideas. By extending these pricing methods to collateralized loans, quantitative trading desks have been able to create a complete new category of financial products that are now used by most banks and other financial institutions.

## Subprime Mortgage Originations

*In 2006, $600 billion of subprime loans were originated, most of which were securitized. That year, subprime lending accounted for 23.5% of all mortgage originations.*

IN BILLIONS OF DOLLARS

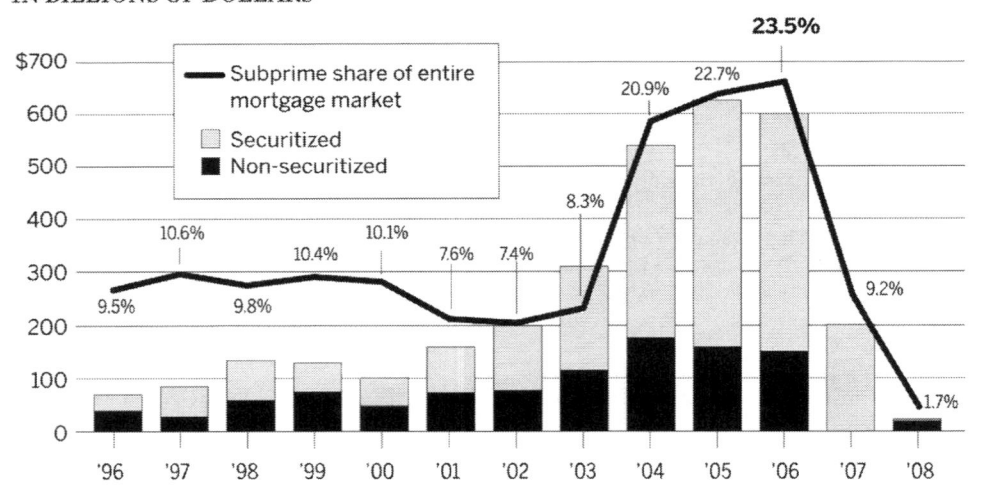

NOTE: Percent securitized is defined as subprime securities issued divided by originations in a given year. In 2007, securities issued exceeded originations.

SOURCE: Inside Mortgage Finance

*Figure 2-2.* Securitization levels of mortgage loans during the 90s and 2000s (from the official government publication, "Financial Crisis Inquiry Commission Report")

# FX Derivatives

Derivatives based on foreign currencies are a relatively simple extension of the ideas already used on options. The underlying price is defined by foreign exchanges. The basic difference between such products and standard options is that they depend on the price variation of two currencies.

FX derivatives have an important role in markets that rely on foreign trade. For example, it is used in the production planning of companies that need protection against variations in currency prices. Most global companies that buy or sell products in a foreign market will use FX derivatives as a tool to avoid the unpredictability of currency fluctuations.

FX derivatives can also be an investment vehicle. Hedge funds have for a long time used foreign exchange products as a way to hedge against possible losses in foreign investments. They can also be used to speculate on the rise or fall of foreign currencies as compared to the local currencies. For all these reasons, the pace of development of mathematical models for FX derivatives has been significant in the industry. Because of the right volatility and near real-time needs of FX traders, C++ has become the language of choice for developing applications that handle FX derivative pricing.

# Derivative Modeling Equations

The equations that have been used to model the future price of derivatives are generally called the Black-Scholes-Merton equations. These equations, which are based on similar differential equations from physics, describe the properties of pricing models when considering a number of input parameters. Here are the most commonly used parameters for these differential equations:

- *The price of the underlying asset:* This is the price of the asset that is the basis for the derivative. In the case of stock options, this is the price of the stock at the present time.

- *The current interest rates:* Interest rates have an important role in the modeling of derivatives, because they are the safest way to get a return on your money. The price of a derivative has to take in consideration the prevailing interest rates and the money that the investor could be earning in a risk-free investment.

- *The strike price:* The price at which a transfer of value will happen. For call options, this is the price above which a profit is made. More complex models will have different definitions for the strike price.

- *Volatility:* The volatility for the underlying asset is very important in derivative models, because it determines how fast the underlying prices move. This information then can be used to calculate the probabilities that are part of the general model for the derivative price. Volatilities are described in terms of the standard deviation.

- *Time left in the contract:* Time is another important variable, because the more time that's left to expiration, the higher the probability that the underlying asset will move in price. This directly affects the probability of profit for the derivative.

These parameters are used as part of the differential equation that determines the price of a derivative. Here is the basic equation that is generally called the Black-Scholes model:

$$\frac{\partial V}{\partial t} + \frac{1}{2}\sigma^2 S^2 \frac{\partial^2 V}{\partial S^2} + rS\frac{\partial V}{\partial S} = rV$$

The differential equation determines the relationship between the following quantities:

- $V$: The price of the derivative
- $t$: The time
- $\sigma$: The volatility
- $S$: The price of the underlying asset
- $r$: The current interest rate

This equation can be extended for different purposes, depending on the type of contract you want to price. For example, when working with options, this equation will result in a formula that returns the price of a call or put option, which will depend on the desired strike. Moreover, the exact formula will change depending on the type of exercise: either an American- or European-style option. You will see in later chapters a few examples of how the general equation can be used with different derivatives.

## Numerical Models

As discussed in the previous section, the existing models for options pricing are based on the Black-Scholes equation, which describes the variation of derivative prices with time, along with a number of other parameters. Later, Merton successfully expanded this model to deal with other derivatives. All these models share the fact that prices are assumed to be random and change according to a predefined probability distribution.

In order to solve these models, you have to develop a few techniques to extract the desired prices, given the set of input parameters required by the equations. There are two main strategies that have been devised for this purpose: numerical methods and simulation methods.

Numerical methods refer to a set of mathematical and computational techniques to solve, or at least approximate, differential equations. While numerical methods were invented to solve problems in physics and engineering, they have been recently used with success to solve pricing problems for options and derivatives. Many of the techniques studied in this book are targeted at solving one or more parts of the derivative-pricing models previously described.

Examples of mathematical tools that are used in the numerical solution of complex derivative models include linear algebra, optimization, and approximation methods, probability, numerical root calculation, and finite difference methods. These mathematical tools can be used in isolation or combined to form more complex algorithms for the solution of Black-Scholes equations.

The other side of numerical models is the development of fast algorithms. While the mathematical tools are important, they need to be implemented in a fast and efficient manner to be used in financial applications. Pricing models frequently need to be solved very frequently, and the performance and accuracy of solutions can make the difference between a profitable and a losing financial transaction.

## Binomial Trees

Another technique used to determine the price of derivatives is the method of binomial trees. A binomial tree is a technique to organize the computation necessary to determine derivative prices in a step-by-step fashion. The root of the tree is the original price. At each node, there are two possible directions for the new price, which can be calculated using a few equations.

Once the complete tree has been calculated, it is possible to answer questions about the fair price of the derivative at particular strike prices and time periods. The complete algorithm for binomial trees has three main steps:

- *The forward phase:* This phase is where the tree is constructed, starting at time zero with the initial price. Then, the total time is divided into discrete steps and at each step a new set of nodes is created. The nodes represent the two directions in which the underlying price can change. This phase starts when the tree nodes reach the maturity date.

- *The payoff phase:* In this phase the profit (payout) of each node is calculated. The calculation starts from the maturity date, since the profit in that case is easy to calculate.

- *The backward phase:* In this phase, the computation of the payout continues moving backwards in time, using the values calculated in the previous phase as the starting point. This process continues until the initial node is reached.

# Simulation Models

Simulation models, also called Monte Carlo models, are a different approach to solve problems involving differential equations, such as the equations necessary for derivative pricing. The main motivation behind simulation models is that the equations for derivative pricing generally don't have a closed mathematical solution. In that case, a possible strategy is to run a simulation of the price evolution, while considering that price changes according to the random distribution assumed by the Black-Scholes equations.

Monte Carlo methods have a long history. Since the development of probability theory, researchers have found that simulating a random event is a good way to learn about a certain physical or engineering model. With the introduction of modern computers, it is now possible to perform very complex simulation in an efficient way. This is an area where using C++ is a big advantage, since simulation accuracy is directly related to the number of repetitions of a basic random experiment.

To find the price of a derivative security, the basic step is to develop a random walk model for the security. As discussed previously, derivatives are based on the idea that underlying prices are always moving in an unpredictable, random way. A Monte Carlo algorithm will use this property to simulate the movements of the underlying asset for a large number of times. The random fluctuations are determined with a random number generator, according to the parameters that have been previously observed for the asset, such as volatility, current interest rate, and observed price of the underlying instrument.

If the simulation is properly performed, a Monte Carlo algorithm will converge to a particular value of derivative price, according to the assumptions of the Black-Scholes equation. The interpretation of these simulated runs can be used to determine the price of a particular contract.

Another consideration is that numeric and Monte Carlo methods are not necessarily exclusive options. You can code numerical methods to solve a particular pricing problem, while at the same time using Monte Carlo method for confirmation of the results. You can also start using Monte Carlo methods to explore different scenarios, and then code a more precise numerical algorithm to find the solution of the more interesting scenarios. Still another possibility is to use numerical algorithms to solve particular sub-problems, and use a Monte Carlo simulation to put these values together in a more complicated scenario. In summary, there are many ways to combine numerical algorithms and simulation to achieve the desired results.

CHAPTER 2 ■ FINANCIAL DERIVATIVES

# Using the STL

One of the main goals of C++ is to provide efficient and high-level code for applications. One of the tools used by programmers to achieve this goal is the Standard Template Library (STL). With the STL, it is possible to create fast containers for generic objects, without incurring runtime inefficiencies.

The STL provides a list of software components that you can use in several contexts. The library can be described as having three main groups of templates:

- *Containers*: A container is a template that provides generic logic for a group of objects. They typically implement traditional data structures using the facilities provided by templates. Table 2-2 displays a quick list of containers provided with the STL and a short description of each one.

- *Iterators*: Along with containers, you also need data structures to manipulate data. This is possible in the STL by using iterators. With an iterator, you can easily access individual elements in a container and perform common operations such as adding, removing, and modifying single elements.

- *Algorithms*: The last major piece of the STL is a set of algorithms that have been optimized to each container. Because templates are parameterized, the algorithms in the STL can be specialized for each container, so that users can have the fastest algorithm for each container while using the same interface. This means that you just need to learn a small set of algorithms that are applicable to several containers. The STL templates will guarantee that you're using the most efficient version for that particular container. Table 2-3 displays a quick list of algorithms in the STL.

*Table 2-2. List of STL Containers*

| Container | Description |
|---|---|
| std::vector | A dynamically allocated array of elements, where members are guaranteed to be allocated contiguously. |
| std::list | A linked list data structure. |
| std::map | An associative data structure, where elements are associated with keys of a particular type. |
| std::multimap | A version of std::map template that can also contain repeated elements. |
| std::queue | A first-in last-out template. |
| std::dqueue | A double queue, where elements can be added or removed from both sides of the queue. |
| std::set | A data structure that contains ordered values and provides quick lookup functionalities. |
| std::multiset | A set where elements can appear more than once. |

*Table 2-3.* *List of STL Algorithms*

| Algorithm | Description |
| --- | --- |
| std::for_each | Performs a given function for each element of the target container. |
| std::find | Searches the container for a given element, given a range indicating the beginning and end of the region. |
| std::find_if | Similar to std::find, but searches the container for a given element satisfying a given predicate. |
| std::find_first_of | Searches the container for the first match of a particular element, given a range of elements. |
| std::count | Counts the number of elements in the container that matches the given parameter. |
| std::count_if | Counts the number of elements in the container that satisfies a given predicate. |
| std::copy | Copies elements from a given origin position to a destination. |
| std::move | Moves elements from a given origin position to a destination position. |
| std::reverse | Reverses the current order of the container. |
| std::sort | Sorts the container according to a comparison function. |
| std::binary_search | Performs binary search for a particular element on a given container. |

## Generating a Random Walk

This section describes a simple way to generate random walks in C++. While the method presented is not optimal, it shows most of the elements necessary to create realistic random walks. In the later chapters, you learn about the statistical techniques that can be used to create more realistic random walks, suitable for derivative pricing algorithms.

The class is called RandomWalkGenerator and it exposes a main member function called generateWalk(). The class has the responsibility of creating a sequence of numbers that represent a random walk. This means that starting on a particular value (the initial price), the sequence will change according to random increments, as determined by the given step parameter. Finally, the size of the sequence (which corresponds to the time to expiration of a contract) is also given as a parameter to the class. This results in a class with the following signature to the constructor:

```
RandomWalkGenerator(int size, double start, double step);
```

The main member function performs the task of sequentially generating new steps in the price simulation. The function receives no parameters and uses the member data already stored in the RandomWalkGenerator class.

The way this member function operates is based on the std::vector container, which is used to store all the intermediate prices created by the Monte Carlo process. The constructor used in this case is the default constructor, which results in an empty vector. The vector is then populated inside the for loop using the vector::push_back, a member function that adds a new element at the end of the vector and resizes the vector if more space is necessary. The fragment uses the value returned by computeRandomStep(), starting from the previous price stored in the variable prev.

```
std::vector<double> RandomWalkGenerator::generateWalk()
{
    vector<double> walk;
    double prev = m_initialPrice;

    for (int i=0; i<m_numSteps; ++i)
    {
        double val = computeRandomStep(prev);
        walk.push_back(val);
        prev = val;
    }
    return walk;
}
```

Finally, you can see the computeRandomStep member function, which generates a new random price according to the simulation arguments. The idea used in this example is that there is a 1/3 chance that the price will change up, down, or stay the same. I use a simple random number generator to return uniformly generated numbers (rand is not the best choice for such applications, but you'll learn about better options in a latter chapter). The result is that you have a "three-sided coin" that determines the direction of the next step in the simulation. Here is the complete code for this member function:

```
double RandomWalkGenerator::computeRandomStep(double currentPrice)
{
    const int num_directions = 3;
    int r = rand() % directions;
    double option_value = currentPrice;

    if (r == 0)
    {
        option_value += (m_stepSize * val);
    }
    else if (r == 1)
    {
        option_value -= (m_stepSize * val);
    }
    return option_value;
}
```

Finally, I present a test stub that can be used to verify the code. It is always a great idea to perform some testing of the algorithm as you implement it. This kind of testing can be used to avoid obvious mistakes as you code an algorithm. The test case is to generate a random walk starting from price $30, for 100 steps with a step size of $0.01. Here is the code that's used:

```
int main()
{
    // 100 steps starting at $30
    RandomWalkGenerator rw(100, 30, 0.01);
    vector<double> walk = rw.generateWalk();
```

```
        for (int i=0; i<walk.size(); ++i)
        {
            cout << ", " << i << ", " << walk[i] << cout::endl;
        }
        cout << endl;
        return 0;
}
```

## Complete Listing

The complete code for the example is listed next. The code is split into a header file called GenericOption.h and an implementation file called GenericOption.cpp.

***Listing 2-1.*** Interface of the RandomWalkGenerator class

```
//
//  RandomWalkGenerator.h
//
// Interface for random walk generator class.

#ifndef __CppOptions__RandomWalkGenerator__
#define __CppOptions__RandomWalkGenerator__

// the class uses a vector to hold the elements
// of the random walk, so they can be later plotted.
#include <vector>

//
// Simple random walk generating class. This class can be
// used for price simulation purposes.
//
class RandomWalkGenerator {
public:
    //
    // class constructors
    RandomWalkGenerator(int size, double start, double step);
    RandomWalkGenerator(const RandomWalkGenerator &p);

    // destructor
    ~RandomWalkGenerator();

    // assignment operator
    RandomWalkGenerator &operator=(const RandomWalkGenerator &p);

    // main method that returns a vector with values of the random walk
    std::vector<double> generateWalk();

    // returns a single step of the random walk
    double computeRandomStep(double currentPrice);
```

CHAPTER 2 ■ FINANCIAL DERIVATIVES

```
private:
    int m_numSteps;        // the number of steps
    double m_stepSize;     // size of each step (in percentage points)
    double m_initialPrice;      // starting price
};

#endif /* defined(__CppOptions__RandomWalkGenerator__) */
```

***Listing 2-2.*** Implementation of the RandomWalkGenerator class

```
//
//  RandomWalkGenerator.cpp
//
//  Simple random walk implementation.

#include "RandomWalkGenerator.h"

#include <cstdlib>
#include <iostream>

using std::vector;
using std::cout;
using std::endl;

//
// Constructor. The supplied parameters represent the number of elements in the
// random walk, the initial price, and the step size for the random walk.
//
RandomWalkGenerator::RandomWalkGenerator(int size, double start, double step)
: m_numSteps(size),
m_stepSize(step),
m_initialPrice(start)
{
}

RandomWalkGenerator::RandomWalkGenerator(const RandomWalkGenerator &p)
: m_numSteps(p.m_numSteps),
m_stepSize(p.m_stepSize),
m_initialPrice(p.m_initialPrice)
{
}

RandomWalkGenerator::~RandomWalkGenerator()
{
}

RandomWalkGenerator &RandomWalkGenerator::operator=(const RandomWalkGenerator &p)
{
```

```cpp
        if (this != &p)
        {
            m_numSteps = p.m_numSteps;
            m_stepSize = p.m_stepSize;
            m_initialPrice = p.m_initialPrice;
        }
        return *this;
}

//
// returns a single step of the random walk
//
double RandomWalkGenerator::computeRandomStep(double currentPrice)
{
    const int num_directions = 3;
    int r = rand() % num_directions;
    double val = currentPrice;
    if (r == 0)
    {
        val += (m_stepSize * val);
    }
    else if (r == 1)
    {
        val -= (m_stepSize * val);
    }
    return val;
}

//
// This is the main method. It will generate random numbers within
// the constraints set in the constructor.
//
std::vector<double> RandomWalkGenerator::generateWalk()
{
    vector<double> walk;
    double prev = m_initialPrice;

    for (int i=0; i<m_numSteps; ++i)
    {
        double val =  computeRandomStep(prev);
        walk.push_back(val);
        prev = val;
    }
    return walk;
}

//
// This is a testing stub. It generates a sequence of random points.
//
```

CHAPTER 2 ▪ FINANCIAL DERIVATIVES

```
int main()
{
    // 100 steps starting at $30
    RandomWalkGenerator rw(100, 30, 0.01);
    vector<double> walk = rw.generateWalk();

    for (int i=0; i<walk.size(); ++i)
    {
        cout << ", " << i << ", " << walk[i] << cout::endl;
    }
    cout << endl;
    return 0;
}
```

## Building and Testing

You can build the code presented in the last section using any standards compliant C++ compiler. The code was tested on Linux and MacOS X. You can use a compiler such as gcc, which is freely available on all major platforms. The commands used in this case were:

`gcc -o RandomWalkGenerator.o -c RandomWalkGenerator.cpp`

`gcc -o RandomWalkGenerator RandomWalkGenerator.o`

The code contains a test stub that generates a sample random walk. You can run the application to see the sequence of random prices created by the RandomWalkGenerator class. Here is sample output from my machine:

`$ ./RandomWalkGenerator`

**0, 29.7,**
**1, 29.403,**
**2, 29.403,**
**3, 29.403,**
**4, 29.109,**
**5, 29.109,**
**6, 29.4001,**
**7, 29.4001,**
**8, 29.4001,**
**9, 29.1061,**
**10, 29.3971,**
**...**

Using the data provided in this sample output, it is easy to create a chart that shows the price behavior over a simulated period of time, as shown in Figure 2-3. Notice how this simple output is close to the behavior of a traded asset. You will later learn to change the parameters in this type of simulation so that it more closely resembles a particular asset class.

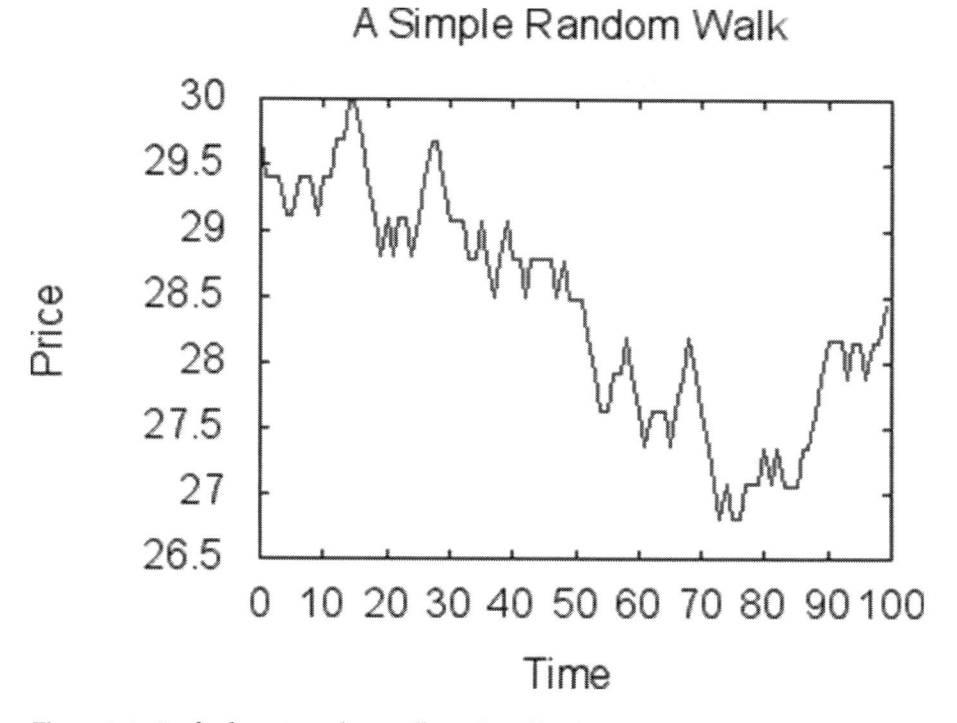

*Figure 2-3.* *Profit chart A random walk produced by the application RandomWalkGenerator*

## Further References

Derivatives are a broad subject, and several books have been written on theoretical and practical aspects of these investment vehicles. Here is a quick list of references that can be used to get additional information on this topic.

- *Practical C++ Financial Programming*, by C. Oliveira. This book covers most of the basic algorithms necessary for derivatives pricing. Examples in C++ are provided in each chapter.

- The "*Financial Crisis Inquiry Commission Report,*" which is a publication of the U.S. government (available at http://www.gpo.gov/fdsys/pkg/GPO-FCIC/pdf/GPO-FCIC.pdf) provides an overview of derivatives trading activity that lead to the financial crisis of 2008.

- *Options, Futures, and Other Derivatives*, by John C. Hull. This is the standard textbook introduction to derivatives.

- *Derivatives Markets*, by Robert L. McDonald. This book provides an in-depth look at the several markets in which financial derivative methods have been applied.

## Conclusion

This chapter introduced the main ideas about general derivatives. Derivatives allow investors and traders to enter into contracts that are based on a particular asset, while having some of their rights defined by associated price levels of the underlying asset and other parameters, such as interest rates, volatility, and time to expiration. The concepts behind derivatives make it possible to create financial products that uniquely target different patterns of risk and reward. Derivatives can be used to mitigate the risk associated with many credit and asset based transactions. They can also be used to make risky bets on particular markets.

You have seen the basic models used for derivative pricing. These models are ultimately based on the equations developed by Black, Scholes, and Merton. These partial derivative equations determine with precision the price of the derivative as time passes, while making a small number of assumptions about the underlying asset. The main assumption used is that the changes in the underlying asset are randomly distributed, with known volatility.

I described the main approaches for solution derivative pricing models. In general terms, you will be able to apply numerical algorithms, based on the exact solution of mathematical equations, binary tree techniques, or Monte Carlo methods, which are simulation algorithms that replicate the price movements of the desired financial asset.

As an example of C++ programming for derivative pricing, I introduced a C++ class that implements a random walk. This class illustrates how Monte Carlo methods operate, and will be later used as a basic algorithm for more complex pricing methods.

The next chapter introduces other basic algorithms used in the implementation of option and derivative pricing models. You will see how these algorithms can be efficiently coded in C++. I also reviewed some of the most used C++ libraries in finance.

# CHAPTER 3

# Basic Algorithms

To become a proficient software developer, it is essential that you understand the basic algorithms used in your application area. This is especially applicable to financial derivatives, where some basic problems and algorithms are recurring. In this chapter, I examine some common algorithms encountered in C++ applications for analyzing and processing options and derivatives.

Some of the basic algorithms in this area involve recurring tasks such as time series processing, date and time handling, and data access and storage. While these algorithms are useful in most applications, they are especially important in code that handles options and other financial derivatives. This chapter will also prepare you for the type of C++ coding skills that are necessary for more advanced topics covered in the following chapters.

The chapter is organized so that you survey some basic algorithms and their implementation in C++, including the following topics:

- *Date and handling:* Date representations are important for many of the underlying algorithms used in financial engineering. You will learn about the main operations performed on dates, and how they can be implemented in C++.

- *Compact date implementation:* Another aspect of date processing is the efficient memory use for long-time series. I discuss some of the alternative representations for date objects and explain how they can be implemented in C++.

- *Networks and graphs:* Data elements and their relationships are often described as a network of connections. This is true for many of the data entities used in financial analysis. You will see a quick overview of networks and their representation using C++ and the STL, along with an example of their use.

## Date and Time Handling

Among the basic algorithms and data structures used on financial algorithms, data handling is one of the most commonly used. Dates are needed to process time series, which can span time periods ranging from a few minutes to several years. For this reason, it is important to use date-handing data structures that are efficient and accurate so that you don't need to worry about the correctness of financial calculations depending on dates.

In this section, you'll learn about the most common ways to represent dates in C++ applications. This will also help you choose a date representation that matches the requirements of your particular application. The first thing is to realize that there are several ways to represent dates in a computer program. The simplest technique is to use a class that stores the values for day, month, and year. This is the representation used for the Date class, as explained in this section. A more compact representation of dates will be presented in the next section.

CHAPTER 3 ■ BASIC ALGORITHMS

# Date Operations

A number of operations are commonly required on dates. Table 3-1 presents some of the most common date operations that will be discussed in this chapter.

***Table 3-1.*** *List of Common Operations Performed on Date Objects*

| Operation | Description |
| --- | --- |
| add | Add a certain number of days to the current date. |
| subtract | Subtract a certain number of days from the current date. |
| addTradingDays | Add a number of trading days to the current date. |
| subtractTradingDays | Subtract a number of trading days from the current date. |
| dateDifference | Return the difference in days from the current date. |
| tradingDateDifference | Return the difference in trading days from the current date. |
| dayOfTheWeek | Return the day of the week corresponding to the current date. |
| isWeekDay | True if the date is a weekday. |
| isHoliday | True if the date is a holiday. |
| isTradingDay | True if the date is a trading day. |
| nextDay | Increment the current date to the next valid day. |
| nextTradingDay | Increment the current date to the next valid trading day. |

The Date class implements the operations described in Table 3-1. The declaration for the Date class is the following:

```
class Date {
public:
    Date(int year, int month, int day);
    Date(const Date &p);
    ~Date();
    Date &operator=(const Date &p);

    void setHolidays(const std::vector<Date> &days);
    std::string month();
    std::string dayOfWeek();

    void add(int numDays);
    void addTradingDays(int numDays);
    void subtract(int numDays);
    void subtractTradingDays(int numDays);
    int dateDifference(const Date &date);
    int tradingDateDifference(const Date &date);
    DayOfTheWeek dayOfTheWeek();
    bool isHoliday();
    bool isWeekDay();
    Date nextTradingDay();
```

```
    bool isLeapYear();
    bool isTradingDay();
    void print();

    Date &operator ++();
    Date &operator --();
    bool operator<(const Date &d) const;
    bool operator==(const Date &d);
private:
    int m_year;
    int m_month;
    int m_day;
    DayOfTheWeek m_weekDay;
    std::vector<Date> m_holidays;
};
```

Notice that the date members store the year, month, and day, which are passed to the constructor. There are two other data members: m_weekDay, which stores the current day of the week (if it is known), and m_holidays, which stores a list of given holidays.

The day of the week is calculated by adding days starting from January 1st, 1900, which was a Monday. This process is improved by storing the result on m_weekDay so that it doesn't need to be recomputed. The member function is implemented as follows:

```
DayOfTheWeek Date::dayOfTheWeek()
{
    if (m_weekDay != DayOfTheWeek_UNKNOWN) return m_weekDay;

    int day = 1;
    Date d(1900, 1, 1);
    for (;d < *this; ++d)
    {
        if (day == 6) day = 0;
        else day++;
    }
    m_weekDay = static_cast<DayOfTheWeek>(day);
    return m_weekDay;
}
```

Another important member function used throughout the class is operator++. This function will update the object so that it represents the next valid date. In most cases, only the m_day field needs to be incremented. However, when the day is 28, 29, 30, or 31, both the month and day need to be updated. Then, the right thing to do depends on the month, as shown in the following code fragment:

```
if (m_day == 31)
{
    m_day = 1;
    m_month++;
}
```

```
        else if (m_day == 30 &&
                std::find(monthsWithThirtyOneDays.begin(),
                          monthsWithThirtyOneDays.end(), m_month)
                    == monthsWithThirtyOneDays.end())
        {
            m_day = 1;
            m_month++;
        }
        else if (m_day == 29 && m_month == 2)
        {
            m_day = 1;
            m_month++;
        }
        // ...
```

Here, monthsWithThirtyOneDays is a vector containing a set of months that have 31 days. Other tests are analogous to this example. Similarly, operator-- adjusts the object to the previous valid date. If the current day is 1, it finds the right date based on the number of days in the previous month.

The isTradingDay member function returns true if the current date is not a holiday or a weekend day:

```
bool Date::isTradingDay()
{
    if (!isWeekDay()) return false;
    if (m_holidays.size() == 0) return true;
    if (isHoliday()) return false;
    return true;
}
```

Most other functions are implemented based on these primitive functions. For example, here is how you can add days to the current date:

```
void Date::add(int numDays)
{
    for (int i=0; i<numDays; ++i)
    {
        ++*this;
    }
}
```

And here is how you can add *trading* days to the current date. First, you find the first trading day starting from the given date. Then, for each trading day add one to the current date and skip all next non-trading days. The implementation is as follows:

```
void Date::addTradingDays(int numDays)
{
    while (!isTradingDay())
    {
        ++*this;
    }
```

```
        for (int i=0; i<numDays; ++i)
        {
            ++*this;
            while (!isTradingDay())
            {
                ++*this;
            }
        }
    }
}
```

## Complete Listings

Here you can find the complete code for the Date class. Listing 3-1 contains the header file and Listing 3-2 shows the implementation file for Date.

*Listing 3-1.* Interface of the Date Class

```
//
// Date.h

#ifndef __CppOptions__Date__
#define __CppOptions__Date__

#include <vector>

enum DayOfTheWeek {
    DayOfTheWeek_Sunday,
    DayOfTheWeek_Monday,
    DayOfTheWeek_Tuesday,
    DayOfTheWeek_Wednesday,
    DayOfTheWeek_Thursday,
    DayOfTheWeek_Friday,
    DayOfTheWeek_Saturday,
    DayOfTheWeek_UNKNOWN
};

enum Month {
    Month_January = 1,
    Month_February,
    Month_March,
    Month_April,
    Month_May,
    Month_June,
    Month_July,
    Month_August,
    Month_September,
    Month_October,
    Month_November,
    Month_December,
};
```

CHAPTER 3 ■ BASIC ALGORITHMS

```cpp
class Date {
public:
    Date(int year, int month, int day);
    Date(const Date &p);
    ~Date();
    Date &operator=(const Date &p);

    void setHolidays(const std::vector<Date> &days);
    std::string month();
    std::string dayOfWeek();

    void add(int numDays);
    void addTradingDays(int numDays);
    void subtract(int numDays);
    void subtractTradingDays(int numDays);
    int dateDifference(const Date &date);
    int tradingDateDifference(const Date &date);
    DayOfTheWeek dayOfTheWeek();
    bool isHoliday();
    bool isWeekDay();
    Date nextTradingDay();
    bool isLeapYear();
    bool isTradingDay();
    void print();

    Date &operator ++();
    Date &operator --();
    bool operator<(const Date &d) const;
    bool operator==(const Date &d);
private:
    int m_year;
    int m_month;
    int m_day;
    DayOfTheWeek m_weekDay;
    std::vector<Date> m_holidays;
};

#endif /* defined(__CppOptions__Date__) */
```

***Listing 3-2.*** Implementation File of the Date Class

```cpp
//
//  Date.cpp
//  CppOptions

#include "Date.h"

#include <string>
#include <iostream>

using std::cout;
using std::endl;
```

```cpp
using std::string;

Date::Date(int year, int month, int day)
: m_year(year),
  m_month(month),
  m_day(day),
  m_weekDay(DayOfTheWeek_UNKNOWN)
{
}

Date::~Date()
{
}

Date::Date(const Date &p)
: m_year(p.m_year),
  m_month(p.m_month),
  m_day(p.m_day),
  m_weekDay(p.m_weekDay),
  m_holidays(p.m_holidays)
{
}

Date &Date::operator=(const Date &p)
{
    if (&p != this)
    {
        m_day = p.m_day;
        m_month = p.m_month;
        m_year = p.m_year;
        m_weekDay = p.m_weekDay;
        m_holidays = p.m_holidays;
    }
    return *this;
}

bool Date::operator<(const Date &d) const
{
    if (m_year < d.m_year) return true;
    if (m_year == d.m_year && m_month < d.m_month) return true;
    if (m_year == d.m_year && m_month == d.m_month && m_day < d.m_day) return true;
    return false;
}

bool Date::operator==(const Date &d)
{
    return d.m_day == m_day && d.m_month == m_month && d.m_year == m_year;
}
```

CHAPTER 3 ▓ BASIC ALGORITHMS

```cpp
void Date::setHolidays(const std::vector<Date> &days)
{
    m_holidays = days;
}

bool Date::isHoliday()
{
    return std::find(m_holidays.begin(), m_holidays.end(), *this) != m_holidays.end();
}

std::string Date::month()
{
    switch (m_month) {
        case Month_January: return "January";
        case Month_February: return "February";
        case Month_March: return "March";
        case Month_April: return "April";
        case Month_May: return "May";
        case Month_June: return "June";
        case Month_July: return "July";
        case Month_August: return "August";
        case Month_September: return "September";
        case Month_October: return "October";
        case Month_November: return "November";
        case Month_December: return "December";
        default: throw std::runtime_error("unknown month");
    }
    return "";
}

#define self this

std::string Date::dayOfWeek()
{
    switch (this->dayOfTheWeek()) {
        case DayOfTheWeek_Sunday: return "Sunday";
        case DayOfTheWeek_Monday: return "Monday";
        case DayOfTheWeek_Tuesday: return "Tuesday";
        case DayOfTheWeek_Wednesday: return "Wednesday";
        case DayOfTheWeek_Thursday: return "Thursday";
        case DayOfTheWeek_Friday: return "Friday";
        case DayOfTheWeek_Saturday: return "Saturday";
        default: throw std::runtime_error("unknown day of week");
    }
}

void Date::add(int numDays)
{
    for (int i=0; i<numDays; ++i)
    {
        ++*this;
    }
}
```

```
void Date::addTradingDays(int numDays)
{
    while (!isTradingDay())
    {
        ++*this;
    }
    for (int i=0; i<numDays; ++i)
    {
        ++*this;
        while (!isTradingDay())
        {
            ++*this;
        }
    }
}
void Date::subtract(int numDays)
{
    for (int i=0; i<numDays; ++i)
    {
        --*this;
    }
}
void Date::subtractTradingDays(int numDays)
{
    while (!isTradingDay())
    {
        --*this;
    }
    for (int i=0; i<numDays; ++i)
    {
        --*this;
        while (!isTradingDay())
        {
            --*this;
        }
    }
}
int Date::dateDifference(const Date &date)
{
    Date d = *this;
    if (d < date)
    {
        int diff=0;
        while (d < date)
        {
            ++d;
            ++diff;
        }
        return diff;
    }
```

## CHAPTER 3 ■ BASIC ALGORITHMS

```cpp
        int diff=0;
        while (date < d)
        {
            --d;
            --diff;
        }
        return diff;
}

int Date::tradingDateDifference(const Date &date)
{
    Date d = *this;
    if (d < date)
    {
        int diff=0;
        while (!d.isTradingDay()) ++d;
        while (d < date)
        {
            ++d;
            ++diff;
            while (!d.isTradingDay()) ++d;
        }
        return diff;
    }

    int diff=0;
    while (!d.isTradingDay()) --d;
    while (date < d)
    {
        --d;
        --diff;
        while (!d.isTradingDay()) --d;
    }
    return diff;
}

DayOfTheWeek Date::dayOfTheWeek()
{
    if (m_weekDay != DayOfTheWeek_UNKNOWN) return m_weekDay;

    int day = 1;
    Date d(1900, 1, 1);
    for (;d < *this; ++d)
    {
        if (day == 6) day = 0;
        else day++;
    }
    m_weekDay = static_cast<DayOfTheWeek>(day);
    return m_weekDay;
}
```

```cpp
bool Date::isWeekDay()
{
    DayOfTheWeek dayOfWeek = dayOfWeek();
    if (dayOfWeek == DayOfTheWeek_Sunday || dayOfWeek == DayOfTheWeek_Saturday)
    {
        return false;
    }
    return true;
}

bool Date::isTradingDay()
{
    if (!isWeekDay()) return false;
    if (m_holidays.size() == 0) return true;
    if (isHoliday()) return false;
    return true;
}

Date Date::nextTradingDay()
{
    Date d = *this;
    if (d.isTradingDay())
    {
        return ++d;
    }
    while (!d.isTradingDay())
    {
        ++d;
    }
    return d;
}

bool Date::isLeapYear()
{
    if (m_year % 4 != 0) return false;
    if (m_year % 100 != 0) return true;
    if (m_year % 400 != 0) return false;
    return true;
}

Date &Date::operator--()
{
    if (m_weekDay != DayOfTheWeek_UNKNOWN) // update weekday
    {
        if (m_weekDay == DayOfTheWeek_Sunday)
            m_weekDay = DayOfTheWeek_Saturday;
        else
            m_weekDay = static_cast<DayOfTheWeek>(m_weekDay - 1);
    }
```

```cpp
    if (m_day > 1)
    {
        m_day--;
        return *this;
    }

    if (m_month == Month_January)
    {
        m_month = Month_December;
        m_day = 31;
        m_year--;
        return *this;
    }

    m_month--;

    if (m_month == Month_February)
    {
        m_day = isLeapYear() ? 29 : 28;
        return *this;
    }

    // list of months with 31 days
    std::vector<int> monthsWithThirtyOneDays = { 1, 3, 5, 7, 8, 10, 12 };
    if (std::find(monthsWithThirtyOneDays.begin(),
                  monthsWithThirtyOneDays.end(), m_month)
             != monthsWithThirtyOneDays.end())
    {
        m_day = 31;
    }
    else
    {
        m_day = 30;
    }
    return *this;
}

Date &Date::operator++()
{
    // list of months with 31 days
    std::vector<int> monthsWithThirtyOneDays = { 1, 3, 5, 7, 8, 10, 12 };

    if (m_day == 31)
    {
        m_day = 1;
        m_month++;
    }
```

```cpp
        else if (m_day == 30 &&
                std::find(monthsWithThirtyOneDays.begin(),
                        monthsWithThirtyOneDays.end(), m_month)
                    == monthsWithThirtyOneDays.end())
        {
            m_day = 1;
            m_month++;
        }
        else if (m_day == 29 && m_month == 2)
        {
            m_day = 1;
            m_month++;
        }
        else if (m_day == 28 && m_month == 2  && !isLeapYear())
        {
            m_day = 1;
            m_month++;
        }
        else
        {
            m_day++;
        }

        if (m_month > 12)
        {
            m_month = 1;
            m_year++;
        }

        if (m_weekDay != DayOfTheWeek_UNKNOWN) // update weekday
        {
            if (m_weekDay == DayOfTheWeek_Saturday)
                m_weekDay = DayOfTheWeek_Sunday;
            else
                m_weekDay = static_cast<DayOfTheWeek>(m_weekDay + 1);
        }
        return *this;
}

void Date::print()
{
    cout << m_year << "/" << m_month << "/" << m_day << endl;
}
```

```
int main()
{
    Date d(2015, 9, 12);
    DayOfTheWeek wd = d.dayOfTheWeek();
    cout << " day of the week: " << wd << " " << d.dayOfWeek() <<    endl;
    d.print();

    d.add(25);
    d.print();

    d.addTradingDays(120);
    d.print();
    cout << " day of the week: " << d.dayOfTheWeek() << " " << d.dayOfWeek() <<    endl;

    return 0;
}
```

# A Compact Date Representation

While the Date class presented in the previous section is an adequate implementation of the concept of dates in C++, it still may not be perfect for all applications. One problem with it is that you need to use integers to store each of the different parts of the date, which includes year, month, and day. In the common 64-bit CPU, this takes 24 bytes, which is lot of space for such a small piece of information.

There are a few ways that you can improve the memory use for Date objects. In this section I explain how to do this using a simple format for date storage that uses a character string. If you use four bytes for the year and two bytes for the month as well as the day, the required memory is reduced to just 8 bytes. This format is also commonly used as a date stamp in several applications, so it is easy to verify the correctness of a particular date.

To show how this implementation works, I created a new class called DateCompact, which is a compact representation of Date objects. I only present a few of the operations required from this data type, but you can implement all other methods provided in the Date class using the underlying representation provided by DateCompact.

The only date member of class DateCompact is a string, declared as

```
char m_date[8];
```

Dates are stored using the following member functions:

```
void setYear(int y);
void setMonth(int m);
void setDay(int d);
```

These dates can be retrieved using three corresponding methods:

```
int year();
int month();
int day();
```

For example, to store the year, you just need to convert the given number into a four-character string:

```
void DateCompact::setYear(int year)
{
    m_date[3] = '0' + (year % 10);  year /= 10;
    m_date[2] = '0' + (year % 10);  year /= 10;
    m_date[1] = '0' + (year % 10);  year /= 10;
    m_date[0] = '0' + (year % 10);
}
```

You need to add each number to the character '0' so that the resulting string is printable. The reverse process is easy, you just need to add the characters in the right way:

```
int DateCompact::year()
{
    // (x - '0')  computes the numeric value corresponding to the each character.
    return  1000 * (m_date[0] - '0') + 100 * (m_date[1] - '0')
          + 10 * (m_date[2] - '0') +  (m_date[3] - '0');
}
```

The comparison operators can be easily implemented with the help of the strncmp function from the C string library. The function strncmp returns a negative number if the first argument is lexicographically less than the first, a positive number if the first argument is greater than the second, and 0 if the two strings are equal. For example, the equality operator can be implemented as follows:

```
bool DateCompact::operator==(const DateCompact &d) const
{
    return strncmp(m_date, d.m_date, 8) == 0;
}
```

Similarly, the less than operator has the following implementation:

```
bool DateCompact::operator<(const DateCompact &d) const
{
    // strcmp returns negative values if the first argument is less than the second.
    return strncmp(m_date, d.m_date, 8) < 0;
}
```

## Complete Listings

The full code for the DateCompact class, described in the previous section, is presented in Listings 3-3 and 3-4.

***Listing 3-3.*** Interface of the DateCompact Class

```
//
// DateCompact.h

#ifndef __CppOptions__DateCompact__
#define __CppOptions__DateCompact__
```

```
//
// a compact representation for dates, using a character string
//
class DateCompact {
public:
    DateCompact(int year, int month, int day);
    DateCompact(const DateCompact &p);
    ~DateCompact();
    DateCompact &operator=(const DateCompact &p);

    void setYear(int y);
    void setMonth(int m);
    void setDay(int d);

    int year();
    int month();
    int day();

    void print();

    bool operator==(const DateCompact &d) const;
    bool operator<(const DateCompact &d) const;

private:
    char m_date[8];
};

#endif /* defined(__CppOptions__DateCompact__) */
```

**Listing 3-4.** Implementation of the DateCompact Class

```
//
//  DateCompact.cpp
//
//  Implementation for the DateCompact class

#include "DateCompact.h"

#include <cstring>
#include <iostream>

using std::cout;
using std::endl;

DateCompact::DateCompact(int year, int month, int day)
{
    setYear(year);
    setMonth(month);
    setDay(day);
}
```

```cpp
DateCompact::DateCompact(const DateCompact &p)
{
    strcpy(m_date, p.m_date);
}

DateCompact::~DateCompact()
{
}

DateCompact &DateCompact::operator=(const DateCompact &p)
{
    if (&p != this)
    {
        strcpy(m_date, p.m_date);
    }
    return *this;
}

//
// Use string comparison to determine if the dates are equal
//
bool DateCompact::operator==(const DateCompact &d) const
{
    return strncmp(m_date, d.m_date, 8) == 0;
}

// Use the strncmp function to determine if a date is less than the other.
bool DateCompact::operator<(const DateCompact &d) const
{
    // strcmp returns negative values if the first argument is less than the second.
    return strncmp(m_date, d.m_date, 8) < 0;
}

//
// Functions to calculate the year, month, and days as integers,
// based on the characters contained in the string 'm_date'.
//

int DateCompact::year()
{
    // (x - '0')  computes the numeric value corresponding to the each character.
    return   1000 * (m_date[0] - '0') + 100 * (m_date[1] - '0')
           + 10 * (m_date[2] - '0') +  (m_date[3] - '0');
}

int DateCompact::month()
{
    return  10 * (m_date[4] - '0') +  (m_date[5] - '0');
}
```

CHAPTER 3 ■ BASIC ALGORITHMS

```
int DateCompact::day()
{
    return  10 * (m_date[6] - '0') +  (m_date[7] - '0');
}

void DateCompact::print()
{
    // copy the m_date string into a NULL terminated string (with 9 characters).
    char s[9];
    strncpy(s, m_date, 8);
    s[8] = '\0';              // properly terminate the string
    cout << s << endl;
}

//
// calculate the string corresponding to the given numeric parameter.
//

void DateCompact::setYear(int year)
{
    m_date[3] = '0' + (year % 10);
    year /= 10;
    m_date[2] = '0' + (year % 10);
    year /= 10;
    m_date[1] = '0' + (year % 10);
    year /= 10;
    m_date[0] = '0' + (year % 10);
}

void DateCompact::setMonth(int month)
{
    m_date[5] = '0' + (month % 10);   month /= 10;
    m_date[4] = '0' + (month % 10);   month /= 10;

}

void DateCompact::setDay(int day)
{
    m_date[7] = '0' + (day % 10);   day /= 10;
    m_date[6] = '0' + (day % 10);   day /= 10;
}

#include "Date.h"

int main()
{
    DateCompact d(2008, 3, 17);
    DateCompact e(2008, 5, 11);
    cout << " size of DateCompact: " << sizeof(DateCompact) << endl;

    d.print();
    e.print();
```

```
        if (d < e)
        {
            cout << " d is less than e " << endl;
        }
        else
        {
            cout << " d is not less than e " << endl;
        }

        Date date(2008, 3, 17);
        cout << " size of Date: " << sizeof(Date) << endl;

        return 0;
}
```

## Building and Testing

The previous code can be built using any standards-compliant C++ compiler. Here are the commands used to build the application on MacOS X using gcc:

```
gcc -o DateCompact.o -c DateCompact.cpp
gcc -o Date.o -c Date.cpp

gcc -o main  DateCompact.o Date.o
```

The main function provides a quick test of the DateCompact class, which also compares the size of the objects created using DateCompact and Date. Notice how Date occupies much more memory than DateCompact.

**./DateCompact**
**size of DateCompact: 8**
**20080317**
**20080511**
 **d is less than e**
 **size of Date: 48**

# Working with Networks

Network structures commonly appear in many fields of software development. Such networks are ideal for representing the connections between entities such as objects or more abstract concepts. In financial applications, for example, elements of a network may represent stocks or other asset classes. Connections between elements of the network may represent correlation between assets, among other uses. In this section, I give an overview of networks and explain how they can be presented in C++ applications. A particular example demonstrates the way in which such algorithms can be designed and implemented.

The problem presented here is called word production. A word is a sequence of characters, and it can represent stock tickers in a financial application, for example. Therefore, IBM and CAT may be viewed as application-specific words. These elements are then stored in a dictionary of useful words. The production problem determines how a word can be derived from another using a dictionary. For example, the word CAT can be derived from the word CAR by just changing a single letter. Complex string production can be performed by using multiple productions. Therefore, it is possible to connect several elements of a dictionary using a set of links, where each link represents a single word production.

CHAPTER 3 ▪ BASIC ALGORITHMS

In the string-production problem, you are given a starting word and a destination word. You also have a dictionary of words (for example, a set of stock tickers that you may be interested in trading). Then, the goal is to find the shortest set of productions that can connect the initial word to the final word. For a concrete example, consider the dictionary containing the words LOB, DAG, LOG, CAR, DOG, CAT, COB, CAB, and CAG. If you start from the word CAT and end with the word DOG, a possible solution to the problem is this sequence:

CAT, CAG, DAG, and DOG

This is not the only solution, but it has minimum size (three productions). Another candidate solution is:

CAT, CAB, COB, LOB, LOG, DOG

However, this is not the shortest solution. For simplicity, it is assumed that all words in the dictionary have the same size.

## Creating a Dictionary Class

The first step to solve this problem is to find a representation for the Dictionary object. I created a class that stores the set of words using a vector called m_values. Here is the class definition:

```
class Dictionary {
public:
    Dictionary(int wordSize);
    ~Dictionary() {}
    Dictionary &operator=(const Dictionary &p); // not implemented
    //   ...
    void addElement(const std::string &s);
    void buildAdjancencyMatrix();
    bool contains(const std::string &s);
    const std::vector<std::vector<bool> > &adjList();
    int elemPosition(const std::string &s);
    int size() { return (int)m_values.size(); }
    std::string elemAtPos(int i);
private:
    std::vector<std::string> m_values;
    std::map<std::string, int> m_valuePositions;
    std::vector<std::vector<bool> > m_adjacencyList;
    int m_wordSize;
};
```

There are other three member variables. m_wordSize is used to store the size of words in the dictionary. The m_valuePositions and m_adjacentList variables are explained later.

The first step in the implementation is to define member functions that add elements to the dictionary. For example, this is how you add new words:

```
void  Dictionary::addElement(const string &s)
{
    if (s.size() != m_wordSize)
    {
        throw std::runtime_error("invalid string size");
    }
    m_values.push_back(s);
    m_valuePositions[s] = (int)m_values.size() - 1;
    cout << " added " << s << endl;
}
```

You can use member functions in std::vector to interact with the underlying m_values collection. In this case, the function uses push_back to add new words of the right size. Notice that when a word is stored, the position of the word is also stored in a std::map named m_valuePositions.

The member function elementAtPos returns the word stored in a certain position of the m_values vector:

```
string Dictionary::elemAtPos(int i)
{
    return m_values[i];
}
```

The member function contains returns true if a word is already stored in the dictionary. It uses the find member function of std::map, which when given a map m, returns the value associated with the given key when the element is found, or the value m.end() when the element is not in the map.

```
bool Dictionary::contains(const string &s)
{
    return m_valuePositions.find(s) != m_valuePositions.end();
}
```

Another feature of the Dictionary class is that it returns the position of an element that has been stored in the vector m_values. To speed up this process, Dictionary uses std::map m_valuePositions, which maps between strings and their respective positions. Using this map, it is possible to define the member function elemPosition. The implementation is straightforward:

```
int Dictionary::elemPosition(const string &s)
{
    return m_valuePositions[s];
}
```

Finally, the Dictionary class is responsible for building an adjacency matrix. That is, a matrix that stores the connectivity information for the network of words stored in this dictionary. The way this works is that the matrix has the size $n$ by $n$, where $n$ is the number of words stored. The entries $A_{ij}$ in the matrix are true or false, and true means that the words stored at positions $i$ and $j$ differ by just one character.

## CHAPTER 3 ■ BASIC ALGORITHMS

The first thing that you need to do is create the adjacency matrix for the given set of words stored in the dictionary. This is done using the buildAdjacencyMatrix member function:

```
void Dictionary::buildAdjancencyMatrix()
{
    m_adjacencyList.clear();
    int n = (int)m_values.size();
    for (int i=0; i<n; ++i)
    {
        m_adjacencyList.push_back(vector<bool>(n));
        for (int j=0; j<n; ++j)
        {
            if (diffByOne(m_values[i], m_values[j]))
            {
                m_adjacencyList[i][j] = true;
            }
        }
    }
}
```

The original adjacency data is cleared and a loop is run through each pair of words stored in m_values. Then, the algorithm checks if the words differ by just one character using the diffByOne member function. If that is true, then the algorithm can set the value of the adjacency to true. The diffByOne algorithm is also straightforward:

```
bool diffByOne(const string &a, const string &b)
{
    if (a.size() != b.size()) return false;
    int ndiff = 0;
    for (unsigned i=0; i<a.length(); ++i)
    {
        if (a[i] != b[i]) ndiff++;
    }
    return ndiff == 1;
}
```

You just need to count the number of different characters occurring in both strings. The function returns true only if the number of differences is equal to one.

## Calculating a Shortest Path

The challenging part of this algorithm is to find the shortest path between the two given nodes of the network, represented by the initial and final words. There are a few alternative algorithms to find a shortest path, but this implementation uses the well-known Dijkstra's algorithm. The central idea of this algorithm is to maintain the known distances starting from the initial node. Then, at each iteration you can look for the neighbors of each node and see if at least one can reduce the known shortest path by traversing that node. If that is possible, then the shortest path starting from that node is updated. This process continues until all nodes in the network have been considered.

I present a simple implementation of this algorithm in the `StringProduction` class. The definition of the class is as follows:

```
class StringProduction {
public:
    StringProduction(Dictionary &d);
    StringProduction(const StringProduction &p);
    ~StringProduction();
    StringProduction &operator=(const StringProduction &p);

    bool produces(const std::string &src, const std::string &dest, std::vector<std::string> &path);
    void shortest_path(int v, int dest, int n, std::vector<std::string> &path);
    std::vector<int> recoverPath(int src, int dest, const std::vector<int> &P, std::vector<int> &path);
private:
    Dictionary &m_dic;
};
```

The `StringProduction` class keeps a reference to a dictionary, which contains all the nodes in the network for use by the shortest path algorithm. The central member function for this class is `shortest_path`, which returns the shortest path between the two given words (which should be part of the underlying dictionary). The first part of the function initializes the data structures used:

```
// initialize the set of distances and the set of nodes
for (int i = 0; i <n; i++) {
    Q.insert(i);
    if (i != v) {
        dist[i] = INF;
    }
}
```

The object named Q has type `std::set`, and it can quickly add and remove elements that will later be checked by the algorithm. The loop is just adding all nodes to Q and setting the initial distances in the vector `dist` to a large number (`INF`). The only exception is the distance between the initial node `v` and itself, which is known to be zero.

Another important part of the algorithm is the so-called relaxation step, where the distance is updated to the latest known shortest-path value:

```
        for (int i=0; i<n; ++i){


            if (A[u][i]) {              // this is a neighbor
                int d = dist [u] + 1;
                if (d < dist[i]) {
                    dist[i] = d;
                    prev[i] = u;
                }
            }
        }
    }
```

The vector prev stores the node that is known to be the previous one in the shortest path sequence. The last part of the algorithm is the path-recovery step, where the complete path is retrieved using the information stored in prev:

```
vector<int> npath;
recoverPath(v, dest, prev, npath);
for (unsigned i=0; i<npath.size(); ++i) {
    path.push_back(m_dic.elemAtPos(npath[i]));
}
```

This algorithm uses the member function called recoverPath to find the numeric sequence of nodes used in the shortest path. The for loop then uses that numeric sequence to recover the words from the dictionary. The implementation of the recoverPath method iterates through the previous nodes to construct a sequence:

```
vector<int> StringProduction::recoverPath(
          int src, int dest, const vector<int> &P, vector<int> &path){
    int v = dest;
    while (v != src) {
        path.push_back(v);
        v = P[v];
    }
    path.push_back(src);
    std::reverse(path.begin(),path.end());
    return path;
}
```

Finally, the produces member function uses the algorithm explained previously to find and return the shortest production. First, it checks that the initial and destination words are stored in the dictionary. Then, the function shortest_path is called with the right parameters. The word sequence is returned using the parameter path. The return value is true if there is a valid sequence with a size greater than zero.

```
bool StringProduction::produces(const string &src, const string &dest, vector<string> &path)
{
    if (!m_dic.contains(src) || !m_dic.contains(dest)) return false;

    shortest_path(m_dic.elemPosition(src), m_dic.elemPosition(dest), m_dic.size(), path);

    return path.size() > 0;
}
```

## Complete Listings

Here is the complete listing for the network-based algorithm described in the preceding section. There are five files that contain the full solution. Two files are used for the Dictionary class. Two other files are used for the StringProduction class. Finally, a main file is provided so that you can run a test on the two classes. The files are displayed in Listings 3-5 to 3-9.

CHAPTER 3 ▪ BASIC ALGORITHMS

***Listing 3-5.*** Interface of the Dictionary Class

```
//
//  Dictionary.h

#ifndef __StringProduction__Dictionary__
#define __StringProduction__Dictionary__

#include <string>
#include <vector>
#include <map>

//
// stores the words in the dictionary and provides an adjacency matrix for the words
class Dictionary {
public:
    Dictionary(int wordSize);
    ~Dictionary() {}
    Dictionary &operator=(const Dictionary &p); // not implemented
private:
    Dictionary(const Dictionary &p);            // not implemented
public:
    void addElement(const std::string &s);
    void buildAdjancencyMatrix();
    bool contains(const std::string &s);
    const std::vector<std::vector<bool> > &adjList();
    int elemPosition(const std::string &s);
    int size() { return (int)m_values.size(); }
    std::string elemAtPos(int i);
private:
    std::vector<std::string> m_values;
    std::map<std::string, int> m_valuePositions;
    std::vector<std::vector<bool> > m_adjacencyList;
    int m_wordSize;
};

#endif /* defined(__StringProduction__Dictionary__) */
```

***Listing 3-6.*** Implementation of the Dictionary Class

```
//
//  Dictionary.cpp

#include "Dictionary.h"

#include <iostream>
#include <vector>
#include <map>
#include <set>
#include <queue>
```

59

```cpp
using std::string;
using std::vector;
using std::set;
using std::map;
using std::cout;
using std::endl;
using std::cerr;

Dictionary::Dictionary(int wordSize)
    : m_values(),
m_valuePositions(),
m_adjacencyList(),
m_wordSize(wordSize)
{
}

const std::vector<std::vector<bool> > &Dictionary::adjList()
{
    return m_adjacencyList;
}

Dictionary &Dictionary::operator=(const Dictionary &p)
{
    if (&p != this)
    {
        m_adjacencyList = p.m_adjacencyList;
        m_valuePositions = p.m_valuePositions;
        m_values = p.m_values;
        m_wordSize = p.m_wordSize;
    }
    return *this;
}

//
// true if the words a and b differ by just one character
//
bool diffByOne(const string &a, const string &b)
{
    if (a.size() != b.size()) return false;
    int ndiff = 0;
    for (unsigned i=0; i<a.length(); ++i)
    {
        if (a[i] != b[i]) ndiff++;
    }
    return ndiff == 1;
}

bool Dictionary::contains(const string &s)
{
    return m_valuePositions.find(s) != m_valuePositions.end();
}
```

```cpp
int Dictionary::elemPosition(const string &s)
{
    return m_valuePositions[s];
}

void  Dictionary::addElement(const string &s)
{
    if (s.size() != m_wordSize)
    {
        throw std::runtime_error("invalid string size");
    }
    m_values.push_back(s);
    m_valuePositions[s] = (int)m_values.size() - 1;
    cout << " added " << s << endl;
}

string Dictionary::elemAtPos(int i)
{
    return m_values[i];
}

void Dictionary::buildAdjancencyMatrix()
{
    m_adjacencyList.clear();
    int n = (int)m_values.size();
    for (int i=0; i<n; ++i)
    {
        m_adjacencyList.push_back(vector<bool>(n));
        for (int j=0; j<n; ++j)
        {
            if (diffByOne(m_values[i], m_values[j]))
            {
                m_adjacencyList[i][j] = 1;
            }
        }
    }
}
```

**Listing 3-7.**  Interface of the StringProduction Class

```cpp
//
//  StringProduction.h

#ifndef __StringProduction__StringProduction__
#define __StringProduction__StringProduction__

#include <vector>

class Dictionary;
```

```
class StringProduction {
public:
    StringProduction(Dictionary &d);
    StringProduction(const StringProduction &p);
    ~StringProduction();
    StringProduction &operator=(const StringProduction &p);

    bool produces(const std::string &src, const std::string &dest, std::vector<std::string> &path);
    void shortest_path(int v, int dest, int n, std::vector<std::string> &path);
    std::vector<int> recoverPath(int src, int dest, const std::vector<int> &P, std::vector<int> &path);
private:
    Dictionary &m_dic;
};

#endif /* defined(__StringProduction__StringProduction__) */
```

***Listing 3-8.*** Implementation of the StringProduction Class

```
//
//  StringProduction.cpp

#include "StringProduction.h"

#include "Dictionary.h"

#include <map>
#include <set>

using std::vector;
using std::string;
using std::map;
using std::set;

StringProduction::StringProduction(Dictionary &d)
: m_dic(d)
{
}

StringProduction::StringProduction(const StringProduction &p)
: m_dic(p.m_dic)
{
}

StringProduction::~StringProduction()
{
}
```

```cpp
StringProduction &StringProduction::operator=(const StringProduction &p)
{
    if (&p != this) {
        m_dic = p.m_dic;
    }
    return *this;
}

//
// recovers the path from a list of previous nodes (P)
vector<int> StringProduction::recoverPath(int src, int dest, const vector<int> &P,
vector<int> &path){
    int v = dest;
    while (v != src) {
        path.push_back(v);
        v = P[v];
    }
    path.push_back(src);
    std::reverse(path.begin(),path.end());
    return path;
}

//
// computes the shortest path.
// node v is the source, dest is destination. If the path can be found, it is stored on parameter path
void StringProduction::shortest_path(int v, int dest, int n, vector<string> &path)
{
    const std::vector<std::vector<bool> > &A = m_dic.adjList(); // get the adjacency matrix
    path.clear();

    vector<int> dist(n, 0);
    vector<int> prev(n, 0);
    set<int> Q;                  // set of nodes
    const int INF = INT_MAX; // a large number

    // initialize the set of distances and the set of nodes
    for (int i = 0; i <n; i++) {
        Q.insert(i);
        if (i != v) {
            dist[i] = INF;
        }
    }
```

63

```cpp
        // this is Dijkstra's algorithm
        while (!Q.empty()) {

            int min = INF;
            int u = -1;
            for (set<int>::iterator it = Q.begin(); it != Q.end(); ++it) {
                // find the minimum value in queue
                if (dist[*it] < min) {
                    min = dist[*it];
                    u = *it;
                }
            }

            Q.erase(u);    // remove min vertex u from set

            // relaxation step
            for (int i=0; i<n; ++i){

                if (A[u][i]) {              // this is a neighbor
                    int d = dist [u] + 1;
                    if (d < dist[i]) {
                        dist[i] = d;
                        prev[i] = u;
                    }
                }
            }
        }

        // recover the path from vector prev
        vector<int> npath;
        recoverPath(v, dest, prev, npath);
        for (unsigned i=0; i<npath.size(); ++i) {
            path.push_back(m_dic.elemAtPos(npath[i]));
        }

}

//
// returns true if the word src produces dest using the dictionary dic
// If true, then path will contain the path between src and dest
//
bool StringProduction::produces(const string &src, const string &dest, vector<string> &path)
{

    if (!m_dic.contains(src) || !m_dic.contains(dest)) return false;

    shortest_path(m_dic.elemPosition(src), m_dic.elemPosition(dest), m_dic.size(), path);

    return path.size() > 0;
}
```

*Listing 3-9.* The main Function with a Simple Test for the StringProduction Class

```cpp
//
//  main.cpp
//  StringProduction
//

#include "StringProduction.h"
#include "Dictionary.h"

#include <iostream>

using std::vector;
using std::string;
using std::cout;
using std::endl;

//
// main function is a test case for the algorithm.
//
int main(int argc, const char * argv[]) {
    
    if (argc != 3) {
        cout << "prog word1 word2" << endl;
        return 1;
    }
    
    Dictionary dic(3);
    dic.addElement("lob");
    dic.addElement("dag");
    dic.addElement("log");
    dic.addElement("car");
    dic.addElement("dog");
    dic.addElement("cat");
    dic.addElement("cob");
    dic.addElement("cab");
    dic.addElement("cag");
    
    dic.buildAdjancencyMatrix();
    
    vector<string> path;
    StringProduction sp(dic);
    if (sp.produces(argv[1], argv[2], path)) {
        cout << " -- the first string produces the second" << endl;
        cout << " -- that path has size " << path.size() << ":\n";
        for (unsigned i=0; i<path.size(); ++i) {
            cout << path[i] << "; ";
        }
    } else {
        cout << " the second string does not produce the second" << endl;
    }
    
    return 0;
}
```

## Building and Testing

You can build the code presented in the last section using any standards-compliant C++ compiler. I tested the code on Linux and MacOS X. The commands used to build the project in gcc are the following:

```
gcc -o StringProduction.o -c StringProduction.cpp
gcc -o Dictionary.o -c Dictionary.cpp
gcc -o main.o -c main.cpp
gcc -o StringProduction Dictionary.o StringProduction.o main.o
```

The main function contains test code that creates a new `Dictionary` object, inserts a small set of words, and uses the `StringProduction` class to calculate the shortest path. Here is a sample of the generated output in my system:

```
./StringProduction cat dog

added lob
added dag
added log
added car
added dog
added cat
added cob
added cab
added cag
-- the first string produces the second
-- that path has size 4:
cat; cag; dag; dog;
```

A quick note about the complexity of this algorithm. As explained, the Dijkstra's alogrithm for shortest paths calculation is used. The current implementation uses a matrix of adjacencies, with complexity $O(n^2)$, where $n$ is number of words in the dictionary. This could be improved using more complex implementation schemes (such as adjacency lists and priority queues); however, I decided to use the simplest data structures in order to concentrate on the algorithm itself.

## Conclusion

In this chapter, I presented a few basic algorithms implemented in C++. These algorithms provide examples of how to solve computational problems using C++ and the STL. You read an overview of two interesting problems: date calculation and shortest paths on data networks.

The first sections dealt with date representations and their associated operations. Dates are needed in nearly all financial- and derivative-related applications. They are an intrinsic part of time series for prices, volatility, and other financial information. You saw how to implement commonly used functions to manipulate dates, such as adding and subtracting dates, finding trade dates, and computing date intervals. You also learned how to design a compact date representation, so that only a small amount of memory is necessary to store a large number of date objects.

Finally, I discussed the common problem of implementing a network, with nodes that represent individual data elements and connections between these nodes. I discussed a simple problem based on a dictionary of strings, which can represent stocks of interest, for example. Then, you learned how to create an algorithm that calculates the shortest paths between elements of this basic dataset.

In the next chapter, you will see more examples of using C++ for financial programming. This time, you will learn more about object-oriented techniques, including how they can be used to create high-performance applications to process options and derivative contracts.

# CHAPTER 4

# Object-Oriented Techniques

For the last 30 years, object-oriented techniques have become the standard for software analysis and development. Since C++ fully supports OO programming, it is essential that you have a good understanding of OO techniques in order to solve many of the challenges presented by options and derivatives programming.

This chapter presents a practical summary of the programming topics you need to understand in order to become proficient in the relevant OO concepts and techniques used in the field of options and derivatives analysis. Some of the topics covered in this chapter include:

- *Fundamental OO concepts in C++*: A quick review of OO concepts as implemented in C++, with examples based on derivatives and options.

- *Problem partitioning*: How to partitioning a problem into classes and related OO concepts, using specific C++ techniques.

- *Designing a solution*: How to use classes and objects to solve problems in financial engineering.

- *Reusing OO components*: How to create reusable C++ components that can be integrated to your own full-scale applications, or even distributed as an external library.

## OO Programming Concepts

Object-oriented programming provides set of principles that can facilitate the development of computer software. Using OO programming techniques, you can easily organize your code and create high-level abstractions for application logic and commonly used component libraries. In this way, OO techniques can be used to improve and reuse existing components, as well as simplify the overall development. OO programming promotes a way of creating software that uses logical elements operating at a higher level of abstraction.

CHAPTER 4 ■ OBJECT-ORIENTED TECHNIQUES

When considering different styles of software programming, it is important to use tools and languages that provide an adequate level of support for the desired programming style. C++ was designed to be a multi-paradigm programming language (see Figure 4-1); therefore, it can properly support more than one style of programming, including:

- *Structured programming:* In structured programming, code is organized in terms of functions and data structures. Each function uses standard control flow structures, such as for, while, do, and if/then/else, to organize code. While this programming style was used in isolation, nowadays it is more commonly used as part of an OO or functional approach.

- *Functional programming:* In this style of programming, functions are the most important element of composition. Functions are also used as a first-class citizens: they can be stored and passed as parameters to other functions in this programming paradigm. The C++11 standard has improved support for functional programming, as seen in Chapter 8.

- *Generic, or template-based programming:* Templates allow programmers to create parameterized types. Such types can be used to implement concepts that are independent of the specific type employed. A common example is a container class such as std::vector, which can be used to store values of any type in a sequence of elements stored in contiguous memory.

- *Object-oriented programming:* A programming style where code is organized in classes and shared in the form of objects. In the OO paradigm, objects can respond to operations that are implemented as member functions in C++. Encapsulation and inheritance are common mechanisms used to support the implementation of OO systems.

| Structured Programming | OO Programming | Functional Programming | Template Programming |
| --- | --- | --- | --- |
| Data structures | Classes | First-class functions | Templates |
| Functions | Polymorphism | Closures | Containers |
| Control flow | Encapsulation | Recursion | Generic algorithms |
| for/while/do | Virtual functions | Currying | vector/map/list |
| if/else/switch | Abstract functions | | Metaprogramming |

*Figure 4-1. A comparison of concepts use in four programming paradigms enabled by C++*

C++ offers complete support for OO concepts. Some of these support elements have already been used in the previous chapters of this book, including classes, objects that can be instantiated from these classes, as well as their supporting elements such as constructors, and destructors, among others. In this chapter, you will learn more about OO concepts that are frequently used in real-world applications, with examples that are directly used in the implementation of options and derivatives in C++.

Remember that the main elements of OO programming can be summarized as follows:

- *Encapsulation*: This concept refers to the division of programmatic responsibilities into different language elements. C++ offers classes that can be used to encapsulate desired functionality in a clear way. When planning applications and coding them in C++, it is always a good idea to determine the main concepts that need to be represented as classes and encapsulate the related procedural code into member functions of that class.

- *Inheritance*: C++ allows programmers to extend a class with new operations. This is possible through the concept of inheritance, when a new class assumes all operations previously available in an existing class, called its parent. Inheritance also allows programmers to add new functionality to existing classes, through the inclusion of new member functions that provide the required functionality.

- *Polymorphism*: Inherited classes in C++ extend available classes through the addition of new member functions. Inherited classes also modify the behavior of existing member functions that have been marked with the `virtual` keyword. Polymorphism in C++ is defined through the use of virtual functions, which are then dispatched using a virtual function table, as implemented by most compilers.

Although C++ provides much more than pure OO programming, these elements alone can nonetheless be used to create very complex and efficient applications in various areas, and in this case on financial applications. In the remaining of this chapter, you will see how these OO concepts can be utilized to solve problems occurring on financial derivatives.

---

**Note** Software development using OO techniques not only allows separation between implementation and interface, but it also requires the clear definition of such concepts. A good C++ programmer will excel at decomposing problems into smaller components, which can then be coded into separate classes. While I can only give examples of this process in this book, design and analysis of OO software is a complex and important phase that should be part of your effort during each software project.

---

# Encapsulation

The idea of encapsulation is to define abstract operations that can be implemented by a single class. Once these operations have been made available, clients of a class can use them without being exposed to the internal details of the implementation such as variables, constants, and other internal code that is only used locally to implement the required features.

One of the important aspects of encapsulation is the ability to hide data, which then becomes the member variables of the target class. Consider for example a class that represents a *credit default swap*. The class should contain enough information to determine how to store and trade such financial instruments. For an example of data that must be encapsulated into such a class, you might want to consider the following:

- *Underlying instrument*: The financial instrument that is the basis for the contract. It could be, for example, a set of bonds for a particular company, cash, or some other pre-established financial instrument.

- *Counterpart*: The institution that is the target of the default swap payments. The payment is generally made when the target institution defaults.

CHAPTER 4 ▪ OBJECT-ORIENTED TECHNIQUES

- *Payoff value*: The monetary value of the default swap contract. This payoff is transferred between institutions if the contract payment condition is triggered.

- *Term*: The term of the contract, after which it ceases to exist.

- *Spread cost*: The recurring payment made by the buyer to maintain the contract. Many contracts require equal payments of a spread that is due at regular periods, such as every month or every year.

By using encapsulation to represent a CDS contract, a C++ developer can simply create a class that contains all these data elements. For example, here is a simple CDS class that represents the concepts described previously.

```
enum CDSUnderlying {
    CDSUnderlying_Bond,
    CDSUnderlying_Cash,
    // other values here...
};

class CDSContract {
public:
    CDSContract();
    CDSContract(const CDSContract &p);
    ~CDSContract();
    CDSContract &operator=(const CDSContract &p);

    // other member functions here...

private:
    std::string m_counterpart;
    CDSUnderlying m_underlying;
    double m_payoff;
    int m_term;
    double m_spreadCost;
};
```

With this definition, you encapsulate all the information that corresponds to a CDS contract into a single class. Because the data members are private, this means that only the class can directly access their state. The main advantage of such an arrangement is that no code outside the CDSContract class is allowed to access the private data, achieving true encapsulation.

If it is necessary to provide access to one or more data members of a class, there are two options. The data member could be moved to the public section of the class, but this would make it possible for the data member to change without knowledge of the CDSContract class.

A better way of doing this is to provide an access member function in a case-by-case way. You could, for example, allow the counterpart and payoff member variables to be accessed by other objects through member functions, as shown here:

```
class CDSContract {
public:
    CDSContract();
    CDSContract(const CDSContract &p);
    ~CDSContract();
    CDSContract &operator=(const CDSContract &p);
```

```cpp
    std::string counterpart() { return m_counterpart; }
    void setCounterpart(const std::string &s) { m_counterpart = s;     }
    double payoff() { return m_payoff;    }
    void setPayoff(double payoff) { m_payoff = payoff; }

private:
    std::string m_counterpart;
    CDSUnderlying m_underlying;
    double m_payoff;
    int m_term;
    double m_spreadCost;
};
```

Using this strategy, any change happening to the m_counterpart and m_payoff will occur only through an operation on the CDSContract class. This means that the class can react to any changes in these values, providing proper encapsulation of that data. For example, suppose that you want to reset the payoff value whenever the counterpart for the CDS contracts changes. This could be done the following way:

```cpp
class CDSContract {
public:
    // ...

    std::string counterpart() { return m_counterpart; }
    void setCounterpart(const std::string &s);
    double payoff() { return m_payoff;    }
    void setPayoff(double payoff) { m_payoff = payoff; }

private:
    std::string m_counterpart;
    CDSUnderlying m_underlying;
    double m_payoff;
    int m_term;
    double m_spreadCost;

    static double kStandardPayoff;
};

void CDSContract::setCounterpart(const std::string &s)
{
    m_counterpart = s;
    setPayoff(kStandardPayoff);
}
```

Whenever the counterpart for the contract changes, the class reacts by resetting the payoff to a standard value (defined by the constant kStandardPayoff). That would not be possible if the m_counterpart data member were not properly encapsulated into the CDSContract class.

## Inheritance

The benefits of encapsulation make it easy to implement and maintain code written in C++. However, it is commonly necessary to extend that code to handle situations that could not be anticipated by the designer of the original class. In that case, you can use inheritance as a powerful way to adapt your classes to new requirements.

With the use of inheritance, it is possible to create a new class that contains the same data and behavior as an existing class. The new class is called a *derived class* and the original class is called a *base* or *parent class*. For example, a *loan only credit default swap* is a CDS where the protection is based on secured loans made on the target entity.

This useful type of CDS could be modeled as a new class that inherits from the original CDSContract class. If you need to create a derived class LoanOnlyCDSContract from a base class CDSContract, the C++ syntax would be the following:

```
class LoanOnlyCDSContract : public CDSContract {
public:

    // constructors go here
    void changeLoanSource(const std::string &source);

private:
    std::string m_loanSource;
};
```

The public keyword is used to indicate that the public interface of the base class CDSContract is still available to the new class. The changeLoanSource member function is used to determine the source of the loan used by the CDS contract. The loan source is then stored in the m_loanSource member variable.

Notice that inheritance creates a new class that has access to all of the public and protected interfaces of the base class. So, you still can call any method from the original CDSContract class when working with LoanOnlyCDSContract. On the other hand, private functions and data members are not available to the derived class. If you envision that a class could be used as the base for a hierarchy, it should provide access to some of the non-public interface using protect variables and functions. As a result, inheritance also requires a certain level of cooperation between base and derived classes.

---

**Note** Inheritance requires that the new class be used in a context similar to the original class. Therefore, inheritance shouldn't be used to create classes that have just a superficial similarity to the original class. In particular, a class that inherits from a base class could be used in the same code as the original class. If this is not true for the new class you need, it is better to create a separate class with a specialized interface.

---

Inheritance is the base technology used to accomplish many of the other techniques available in OO programming. Therefore, ideas such as polymorphism and abstract functions are possible due to the use of inheritance.

## Polymorphism

While inheritance in itself provides a useful extension mechanism, its biggest advantage is the possibility of changing the original behavior of the base class in specific situations. In C++, this is enabled by using the virtual keyword to mark member functions that have polymorphic behavior.

CHAPTER 4 ■ OBJECT-ORIENTED TECHNIQUES

For example, suppose that the CDSContract class is required to calculate the contract value at a particular date. This operation can be performed at the class level, but it will be slightly different for each particular implementation. Concrete implementations of the class may want to take into consideration particular factors that are not available at the base class level, such as differences in underlying, contract structures, and calculation models.

For these and other reasons, determining the best way to calculate the contract value may not be possible at the base class, and it must be delegated to derived class. Such derived class will possess additional data that can be used to compute the contract price with more precision than what is possible on the base class.

This behavioral change can be performed in the derived classes if you use C++ virtual mechanism. Syntactically, this polymorphic behavior can be implemented as long as the member function is modified with the virtual keyword in the original class. The virtual keyword is a C++ tool that allows functions to behave differently according to the concrete instance that is executing the function call.

For example, to support the required polymorphic behavior to calculate the contract value, the CDSContract base class should be coded as follows.

```
class CDSContract {
public:
    CDSContract();
    CDSContract(const CDSContract &p);
    ~CDSContract();
    CDSContract &operator=(const CDSContract &p);

    std::string counterpart() { return m_counterpart; }
    void setCounterpart(const std::string &s);
    double payoff() { return m_payoff;    }
    void setPayoff(double payoff) { m_payoff = payoff; }
    virtual double computeCurrentValue(const Date &d);

private:
    std::string m_counterpart;
    CDSUnderlying m_underlying;
    double m_payoff;
    int m_term;
    double m_spreadCost;

    static double kStandardPayoff;
};
```

The virtual double computeCurrentValue(const Date &d); line declares a new member function that can be overridden by derived classes.

■ **Note** Virtual methods need to be recognized by the compiler. Therefore, the virtual keyword has to appear directly in the base class, not only in the derived classes. If a member function is supposed to have polymorphic behavior, you have to use virtual to signal this information to the compiler. Overriding a non-virtual member function doesn't create a polymorphic object and will result in a warning in most compilers.

The classes derived from CDSContract can implement the virtual member function declared previously, so that it can be invoked when instances of that derived class are created. Here how this can be done for the LoanOnlyCDSContract subclass.

The isTradingDay member function returns true if the current date is not a holiday or a weekend day:

```
class LoanOnlyCDSContract : public CDSContract {
public:
    // constructors go here
    void changeLoanSource(const std::string &s);
    virtual double computeCurrentValue(const Date &d);

private:
    std::string m_loanSource;
};
```

The implementation for a virtual function, both in the base class as well as the derived classes, is not different from the syntax used in other member functions. It is used in the compiler to determine the correct way to handle virtual functions that are called.

The use of a virtual function is determined by its polymorphic invocation through pointers and references. For example, consider the following code using CDSContract and LoanOnlyCDSContract:

```
void useContract(bool isLOContract, Date &currentDate)
{
    CDSContract *contract = nullptr;
    if (isLOContract)
    {
        contract = new LoanOnlyCDSContract();
    }
    else
    {
        contract = new CDSContract(); // normal CDS contract
    }

    contract->computeCurrentValue(currentDate);
    delete contract;
}
```

The useContract function is passed two arguments: the Boolean value isLOContract, which indicates that the contract used is a loan-only CDS. The second argument is the current date for use of the contract. The first line in the function:

```
CDSContract *contract = nullptr;
```

determines the base class of the object that will be created. As with any OO object in C++, a pointer (or reference) to a base class can be used to point to objects of any descent class. In this case, a pointer to the CDSContract class (being the base class) can also be used to point to objects of type LoanOnlyCDSContract. The pointer is initialized to nullptr.

CHAPTER 4 ■ OBJECT-ORIENTED TECHNIQUES

> ■ **Note** The keyword **nullptr** was introduced in the C++11 standard. It provides a way to initialize pointers with a null value without the use of a macro such NULL (which is used in C but normally avoided in C++), or the value 0, which can be easily confused with a numeric expression.

The next lines determine the exact type that will be instantiated. If the `isLOContract` flag is set to true, a new object of type `LoanOnlyCDSContract` is created using the new keyword. Otherwise, the function creates an object of type `CDSContract` as the default value. In a more complex application, types should not be encoded using flags, but passed as a parameter or supplied by some of the part of the application.

The next line

```
contract->computeCurrentValue(currentDate);
```

uses the pointer `contract` to perform a polymorphic call to `computeCurrentValue`. The polymorphic call mechanism will determine the correct implementation for the member function, depending on the exact class of the instance pointed to by the `contract` pointer. The next section explains how this mechanism works in practice, and how it affects the creation and use of objects in C++.

## Polymorphism and Virtual Tables

The first step in using polymorphism via virtual functions is to understand how they differ from regular member functions. When a virtual function is called, the compiler has to determine the type of call and translate it into binary code that will perform the call to the correct implementation. This is done in C++ using the so-called *virtual table mechanism*.

A virtual table is a vector of functions that is created for each class that uses at least one virtual function. The virtual table stores the addresses of virtual functions that have been declared for that particular type, as shown in Figure 4-2.

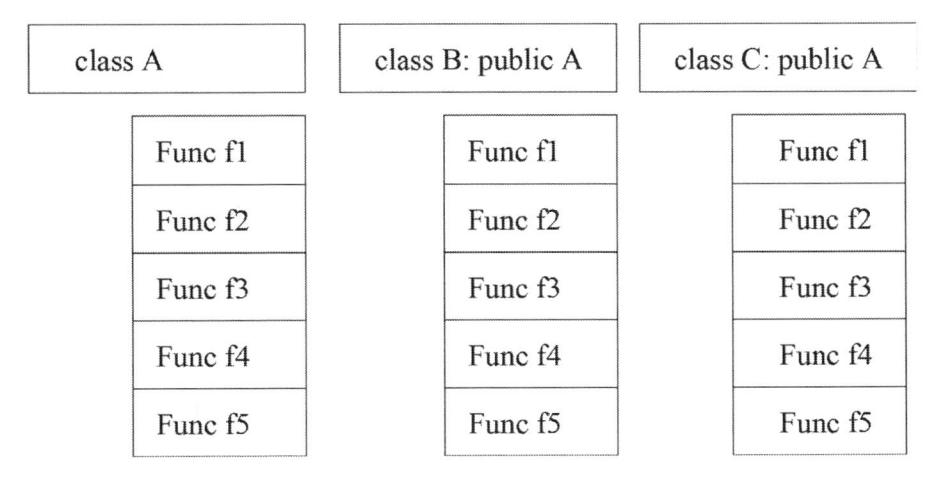

| class A | class B: public A | class C: public A |
|---------|-------------------|-------------------|
| Func f1 | Func f1 | Func f1 |
| Func f2 | Func f2 | Func f2 |
| Func f3 | Func f3 | Func f3 |
| Func f4 | Func f4 | Func f4 |
| Func f5 | Func f5 | Func f5 |

***Figure 4-2.*** *Virtual functions shared by classes A, B, and C and stored in their respective virtual function tables*

As shown in Figure 4-2, class A is the base class and it contains a number of virtual functions, here denoted by the names f1 to f5. The slots in these tables store pointers to the implementation used by the class. Two other classes—B and C—are declared as derived classes via a public interface. This makes classes B and C inherit each a virtual table that contains at least the same function pointers (derived classes can add more virtual functions if they wish to do so).

Each class can define its own version of the virtual function, and as a result the pointer to that function is stored in the corresponding location of the virtual table. The virtual table is populated in the compiler as it creates the data structures necessary for each class. At execution time, the virtual table is available for code executed by each of the classes defined in this example.

During runtime, the code generated by the C++ compiler can retrieve the location in the table where the function pointer is stored. Then the function is called with the given parameters. First, the compiler retrieves the location of the virtual table associated with the class. Then, the compiler finds the function pointer at a predefined displacement from the beginning of the table. Finally, the program makes an indirect call using the function pointer stored at that location.

If you use this information to understand how C++ code works, you can see how the CDSContract and its derived class would execute a call to the computeCurrentValue member function, as shown in the following line of code:

```
contract->computeCurrentValue(currentDate);
```

The first step performed by the implementation is to find the virtual table for the particular object that is stored in the contract pointer. Then, the slot corresponding to the virtual function computeCurrentValue is searched, usually at a fixed distance from the beginning of the vector as determined by the compiler. Finally, the function pointer retrieved in this way is called indirectly, resulting in a function call to the correct implementation.

Although the sequence of steps necessary to call a virtual function appear to be complex, modern compilers can generate very efficient code using the virtual table technique. By means of code optimization, virtual function calls frequently end up as just a call to a function pointer.

## Virtual Functions and Virtual Destructors

Another member function that can be annotated with the virtual keyword is the destructor. As you may remember, a destructor is called automatically (in the code generated by the compiler) when an object goes out of scope, with the objective of reclaiming resources used by the object.

The destructor may also be used through the keyword delete. When a delete is used, the code calls the destructor and frees the memory used by the object up to that moment. As a result, the pointer is not valid after the delete is called.

It is important to consider the role of the destructor when virtual functions are part of a class. The reason is that object cleanup is a class-specific activity, which needs to be overridden for each individual derived class that contains additional resources (such as memory, network connections, or graphical contexts). As a result, the destructor usually has different implementations that are necessary to perform the proper cleanup and de-allocation activities.

For these reasons, the correct way to handle destructors in polymorphic classes is to use the virtual mechanism in their definition. This provides the means for each subclass to call a specific destructor even when called from a base pointer.

CHAPTER 4 ■ OBJECT-ORIENTED TECHNIQUES

For example, consider what happens when the destructor in the base class is not virtual.

```
class CDSContract {
public:
    CDSContract();
    CDSContract(const CDSContract &p);
    ~CDSContract() { std::cout << " base class delete " << std::endl; }
    CDSContract &operator=(const CDSContract &p);

    std::string counterpart() { return m_counterpart; }
    void setCounterpart(const std::string &s);
    double payoff() { return m_payoff;    }
    void setPayoff(double payoff) { m_payoff = payoff; }
    virtual double computeCurrentValue(const Date &d);

// ...
};
```

The derived class LoanOnlyCDSContract would have the following simple definition, which just prints an informational message:

```
class LoanOnlyCDSContract : public CDSContract {
public:
    LoanOnlyCDSContract() { std::cout << " derived class delete " << std::endl; }
    // constructors go here
    void changeLoanSource(const std::string &s);
    virtual double computeCurrentValue(const Date &d);

private:
    std::string m_loanSource;
};
```

If called from client code, these definitions may result in undefined behavior. For example, consider the following fragment:

```
void useBasePtr(CDSContract *contract, Date &currentDate)
{
    contract->computeCurrentValue(currentDate);
    delete contract;
}
```

This code receives a pointer of type CDSContract, uses it to call a virtual function, and then uses the delete operator on it. When called in the following way:

```
void callBasePtr()
{
    Date date(1,1,2010);
    useBasePtr(new LoanOnlyCDSContract(), date);
}
```

77

The code has undefined behavior, because the compiler cannot guarantee that the destructor of the derived class will be found and executed. From the compiler point of view, a non-virtual destructor doesn't need to be called when the object is destroyed.

To fix this problem, the right thing to do is to declare the destructor as virtual in the base class. A simple change in this definition can accomplish this:

```
class CDSContract {
public:
    CDSContract() {}
    CDSContract(const CDSContract &p);
    virtual ~CDSContract() { std::cout << " base delete " << std::endl; }
    CDSContract &operator=(const CDSContract &p);

    // ... other members here

};
```

Once a virtual destructor has been declared in the base class, all descendant classes will also contain a virtual destructor, independent of using the `virtual` keyword. This is guaranteed by the presence of a virtual table containing the address of the destructor, as described in the previous section. The result of the `callBasePtr` function after this change is guaranteed to be the following:

```
$ ./CDSApp
 derived class delete
 base class delete
```

## Abstract Functions

Another mechanism used to implement polymorphism in C++ are abstract functions. Such abstract functions are closely related to virtual functions, but their presence marks the containing class as an abstract class, which cannot be directly instantiated.

An abstract class is frequently used when a function should be provided in derived classes, but there is no clear default behavior that could be provided by the base class. This is a common situation when a base class provides only the framework for an algorithm, with details that are purposefully left unspecified. The idea is that the derived classes will necessarily provide the missing functionality that would make the derived classes useful for a particular application.

The syntax for abstract functions is similar to the syntax for virtual functions. The member function is preceded with the `virtual` keyword as previously seen. In addition, the syntax `= 0;` is used to terminate the declaration of the abstract function. Notice that only a declaration is needed, since no implementation is necessary for an abstract function (although it can be provided if available).

For an example, consider that the `CDSContract` class has a member function to process a credit event. In the world of credit default swaps, a credit event is what happens when a company calls for bankruptcy. Processing this event is different for each entity and CDS type; therefore, I would like to have such a member function as an abstract virtual function:

```
class CDSContract {
public:
    CDSContract() {}
    CDSContract(const CDSContract &p);
    virtual ~CDSContract() { std::cout << " base delete " << std::endl; }
    CDSContract &operator=(const CDSContract &p);
```

```
        std::string counterpart() { return m_counterpart; }
        void setCounterpart(const std::string &s);
        double payoff() { return m_payoff;    }
        void setPayoff(double payoff) { m_payoff = payoff; }
        virtual double computeCurrentValue(const Date &d);

        virtual void processCreditEvent() = 0;

        // ...
};
```

If a base class includes even one abstract virtual function, it becomes an abstract class that cannot be itself instantiated. The reason is that the class can be thought of as "incomplete," since at least one of its virtual functions has no implementation. Given these definitions, the following code would become invalid:

```
CDSContract *createSimpeleContract()
{
    CDSContract *contract = new CDSContract();    /// Wrong: CDSContract is now Abstract
    contract->setCounterpart("IBM");
    return contract;
}
```

Once an abstract member function has been defined, the classes that are direct descents are required to implement that function, or else they will become abstract too. For example, the descendant class LoanOnlyCDSContract now has to implement processCreditEvent in order to be used by client code. Even a trivial implementation would allow LoanOnlyCDSContract to be instantiated.

```
class LoanOnlyCDSContract : public CDSContract {
public:
    LoanOnlyCDSContract() { std::cout << " derived class delete " << std::endl; }
    // constructors go here
    void changeLoanSource(const std::string &s);
    virtual double computeCurrentValue(const Date &d);

    virtual void processCreditEvent();

private:
    std::string m_loanSource;
};

void LoanOnlyCDSContract::processCreditEvent()
{
}
```

Abstract member functions can be freely used even inside the abstract class, where the body of that member function is not defined. For example, this is a valid definition for the CDSContract::computeCurrentValue member function:

```
double CDSContract::computeCurrentValue(const Date &d)
{
    if (!counterpart().empty())
    {
        processCreditEvent(); // make sure there is no credit event;
    }
    return calculateInternalValue();  // use an internal calculation function
}
```

# Building Class Hierarchies

One of the advantages of OO code is the ability to organize your application around conceptual frameworks defined by classes. A class hierarchy allows the sharing of common logic that can be easily reused in other contexts. Proper use of class hierarchy can reduce the amount of code duplication and lead to applications that are more understandable and easy to maintain.

A class hierarchy can be developed around important concepts used by the application. For example, in a derivatives-based application, the class CDSContract would be a candidate to become the root of a class hierarchy. Figure 4-3 shows a possible class hierarchy for CDS contracts, containing derived classes for the following types of contracts:

- LoanOnlyCDSContract: CDS contracts that are based on loans to other institutions and have special logic for processing these loans.

- HedgedCDSContract: A CDS contract type where hedging is performed using other asset classes with the goal of reducing contract risk.

- NakedCDSContract: A particular CDS contract where the contract seller does not own the underlying asset negotiated in the contract.

- FixedInterestCDSContract: A CDS type where the contract requires a fixed interest rate for the duration of the specified agreement.

- VariableInterestCDSContract: A type of CDS where the contracts are defined using variable interest rates, using a well-known benchmark for interest rates.

- TaxAdvantagedCDSContract: A particular type of CDS contract that takes advantage of a special tax structure.

```
                    ┌─────────────────┐
                    │  CDSContract    │
                    └─────────────────┘
                       │           │
         ┌─────────────┘           └─────────────┐
         ▼                                       ▼
┌─────────────────────┐              ┌──────────────────────────┐
│ LoanOnlyCDSContract │              │ FixedInterestCDSContract │
└─────────────────────┘              └──────────────────────────┘

┌─────────────────────┐              ┌──────────────────────────┐
│ HedgedCDSContract   │              │ VariableInterestCDSContract │
└─────────────────────┘              └──────────────────────────┘

┌─────────────────────┐              ┌──────────────────────────┐
│ NakedCDSContract    │              │ TaxAdvantegedCDSContract │
└─────────────────────┘              └──────────────────────────┘
```

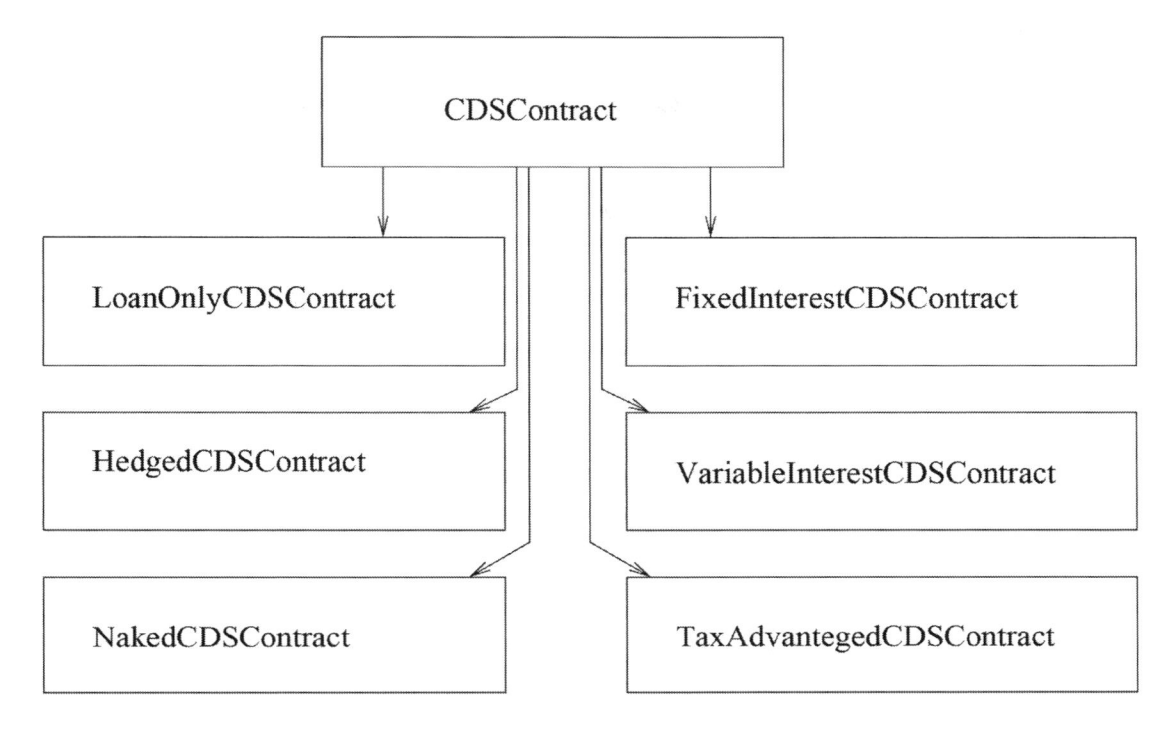

***Figure 4-3.*** *A class hierarchy rooted on the base class CDSContract*

All these CDS contract derivatives would benefit from code sharing from the base class CDSContract. As a result, common functionality such as CDS pricing, contract creation, and contract maintenance can be stored in a central place and used by as many different types of CDS contracts as possible.

Although creating class hierarchies is a useful technique for code maintenance and sharing, inheritance may not be the best strategy for code organization in some cases. It is important to be able to identify the situations in which other approaches would work better. Here are some potential disadvantages of using inheritance:

- *Increased coupling between classes:* Once you decide to use inheritance, there is a big interdependence between classes. A small change in the base class can affect all descendent classes. If there is a situation where the base class can vary frequently in functionality and responsibilities, then inheritance may not be the best solution.

- *Physical dependencies at compilation time:* In C++, inheritance also creates a compile-time dependency between classes. To generate correct code, the C++ compiler needs to access the definition of each base class. This may result in increased compilation time, which is sometimes undesirable, especially in large software projects.

- *Increased information coupling:* Class hierarchies may also require developers to learn the multiple implementations of different classes at different levels. This is necessary especially when classes are not well designed and information about their operations is not clear.

CHAPTER 4 ■ OBJECT-ORIENTED TECHNIQUES

# Object Composition

Another strategy to organize and code using OO techniques is object composition. Composition is an alternative to inheritance, where you can use the behavior of an object without the dependency caused by direct class/subclass relationship.

To use object composition, you need to store the object that has the desired behavior as a member variable for the containing object. This is the basic strategy, which can be implemented in at least three ways in C++:

- *Storing a pointer to an object:* In this case, only a pointer to the object is stored as part of the class. This option allows an object to be created inside the class or passed as a parameter from a user of the class and then stored in a member variable.

- *Storing a reference to an object:* This option allows the class to receive a reference to an existing object, but doesn't allow the object to be created after the constructor is executed. A reference in C++ cannot be reassigned, which leads to a requirement that the stored object needs to be valid the whole time the container object exists.

- *Storing the object as a member variable:* In this case, the containing class assumes responsibility for storing the required object. In this case, it is also necessary for the compiler to know the exact size of the object stored as a member variable.

With object composition, a class can use functionality provided by another class without the use of inheritance.

For example, suppose that the CDSContract class needs a fast calculation of integrals. In this case, a good approach is to use an object-composition strategy to access the functionality of integration, instead of adding this functionality to the base class. You could do this, for example, by passing to the CDSContract constructor a pointer to a MathIntegration object and storing that pointer as a member function. The code would look like this:

```
class MathIntegration;

class CDSContract {
public:
    CDSContract() {}
    CDSContract(MathIntegration *mipt);
    CDSContract(const CDSContract &p);
    virtual ~CDSContract() { std::cout << " base delete " << std::endl; }
    CDSContract &operator=(const CDSContract &p);

    // other member functions here
private:
    std::string m_counterpart;
    CDSUnderlying m_underlying;
    double m_payoff;
    int m_term;
    double m_spreadCost;
    MathIntegration *m_mipt;

    static double kStandardPayoff;
};
```

When necessary, the pointer could be used to access the functionality stored in the `MathIntegration` class. The best thing about this kind of design is that there is little coupling between the `CDSContract` and `MathIntegration` classes. Each one can evolve separately, by adding new functions as necessary, without the need for mutual dependencies.

## Conclusion

In this chapter, you read an overview of OO concepts provided in C++ and how they are used in the financial development community to solve problems occurring with options and derivatives.

The first part of this chapter summarized the basic characteristics of OO as implemented in C++, including the main concepts of encapsulation, inheritance, and polymorphism. You learned about the technique used in C++ to implement polymorphic behavior through virtual functions. You also saw how virtual functions are stored in virtual tables that are created for each class that contains virtual functions.

This chapter also presented some examples of using OO to efficiently solve common problems in financial programming, as applied to options and derivatives. The next chapter proceeds to template-based concepts and explains how they can be used to create high-performance solutions to problems in the area of financial derivatives processing.

# CHAPTER 5

# Design Patterns for Options Processing

Design patterns are a set of common programming design elements that can be used to simplify the solution of recurring problems. With the use of OO techniques, design patterns can be cleanly implemented as a set of classes that work toward the solution of a common goal. These designs can then be reused and shared across applications.

Over the last few years, design patterns have been developed for common problems occurring in several areas of programming. When designing algorithms for options and other derivatives, design patterns can provide solutions that are elegant and reusable (when supporting libraries are employed). Thanks to the inherent ability of the C++ language to create efficient code, these solutions also have high performance.

In this chapter, you will learn about the most common design patterns employed when working with financial options and derivatives, with specific examples of their usage. The chapter covers the following topics:

- *Overview of design patterns*: You will learn how design patterns can help in the development of complex applications, with the ability to reuse common patterns of programming behavior. Using design patterns can also make solutions more robust and easier to understand, because patterns provide a common language that allows developers to discuss complex problems. Such design techniques has also been made available through libraries that implement some of the best known design patterns.

- *Factory method pattern*: A factory method is a design pattern that allows objects to be created in a polymorphic way, so the client doesn't need to know the exact type of the returned object, only the base class that provides the desired interface. It also helps to hide a complex set of creation steps to instantiate particular classes.

- *Singleton pattern*: The singleton pattern is used to model situations in which you know that only one instance of a particular class can validly exist. This is a situation that occurs in several applications, and in finance, I present the example of a clearing house for options trading.

- *Observer pattern*: Another common application of design patterns is in processing financial events such as trades. The observer design patterns allows you to decouple the classes that receive trading transactions from the classes that process the results, which are the observers. Through the observer design pattern, it is possible to simplify the logic and the amount of code necessary to support these common operations, such as the development of a trading ledger, for example.

CHAPTER 5 ■ DESIGN PATTERNS FOR OPTIONS PROCESSING

# Introduction to Design Patterns

Design patterns have been introduced as a set of programming practices that simplify the implementation of common coding problems. As you study the behavior of OO applications, there are tasks and solution strategies that occur frequently and can be captured as a set of reusable classes.

Object-oriented programming provides a set of principles that can facilitate the development of computer software. Using OO programming techniques, you can easily organize your code and create high-level abstractions for application logic and commonly used component libraries. In this way, OO techniques can be used to improve and reuse existing components, as well as simplify the overall development. OO programming promotes a way of creating software that uses logical elements operating at a higher level of abstraction.

Here are some of the most common design patterns that can be used in software development in general and for algorithms to process options and derivatives in particular:

- *Factory method*: In the factory method design pattern, the objective is to hide the complexity and introduce indirection when creating an instance of a particular class. Instead of asking clients to perform the initialization steps, factory methods provide a simple interface that can be called to create the object and return a reference.

- *Singleton*: A singleton is a class that can have at most one active instance. The singleton design pattern is used to control access to this single object and avoid creating copies of this unique instance.

- *Observer*: The observer pattern allows objects to receive notifications for important events occurring in the system. This pattern also reduces the coupling between objects in the system, since the generator of notification events doesn't need to know the details of the observers.

- *Visitor*: The visitor patter allows a member function of an object to be called in response to another dynamic invocation implemented in a separate class. The visitor pattern therefore provides the mechanism for dispatching messages based on a combination of two objects, instead of the single object-based dispatch that is common with OO languages.

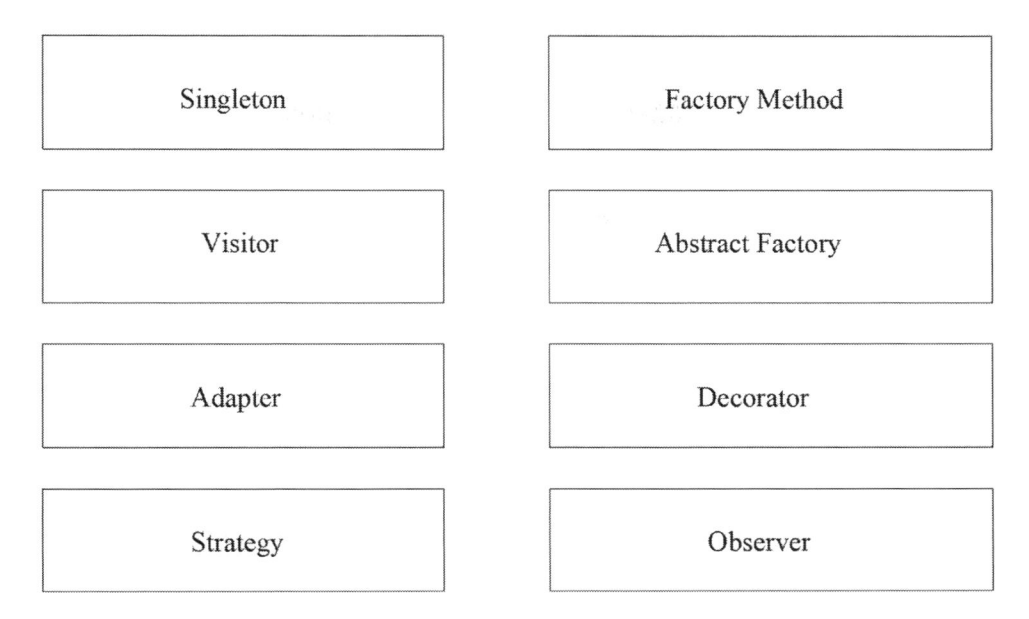

CHAPTER 5 ■ DESIGN PATTERNS FOR OPTIONS PROCESSING

| Singleton | Factory Method |
| --- | --- |
| Visitor | Abstract Factory |
| Adapter | Decorator |
| Strategy | Observer |

*Figure 5-1. A few common design patterns used in OO programming*

In the next few sections, you will see how these design patterns can be implemented in C++, with examples of how they occur in options and derivatives applications.

## The Factory Method Design Pattern

A factory design pattern is a technique used to indirectly create objects of a particular class. This pattern is important because it is frequently useful to access newly allocated objects without having to directly perform the work necessary to create them. For example, using the factory method design pattern, it is possible to avoid the use of the new keyword to create an object, along with the parameters required by the constructor.

The factory design pattern allows an object to be created through a member function of the desired class, so that the client doesn't need to create the object directly. This can be useful for the following reasons:

- Most of the time, there is no need for the client to provide parameters for construction of the object. For example, if the objects require the allocation of additional resources, such as a file or a network connection, the client is relieved from acquiring these resources.

- Sometimes the object depends on internal implementation details, such as a private class, that are not available to clients. In this case, providing a factory method is the only way to create new instances of the object.

- The exact sequence of events necessary to create an object may change. In that case, it is better to provide a factory method that hides this complexity. Users of the class will not have to worry if the way the object is created is updated.

CHAPTER 5 ■ DESIGN PATTERNS FOR OPTIONS PROCESSING

- More importantly, factory methods can be used to simplify *polymorphic object creation*. For example, when an object is created using the new operator, the concrete type of the returned object has to be known by the client. On some applications this might be undesirable, because the real type of the needed object could be any one within a set of derived classes. Using a factory method, it is possible to delegate the creation of the object so that the client code doesn't need to know about the concrete type. As a result, the returned object may be any one of the subtypes of the original type.

Factory methods in C++ are declared as static member functions. Such a member function doesn't depend on a instance of the class to be executed. The syntax for member functions is simply ClassName::functionName(), with parameters added as needed.

■ **Note** The factory method design pattern is also used as a foundation for more complex design patterns. For example, you will notice that other patterns such as singleton use a factory method to control the creation of new instances of a particular class.

In options and derivatives applications, the factory method is commonly used. A situation where the use of a factory method is desirable is when you need to load data objects. The data source used can vary from a local file to a URL, and the parsing of that data is not an important part of the overall algorithm. In that case, abstracting the creation of the data source can be an important application of the factory method.

In the example that follows, you can see how a DataSource class can be implemented. The goal of this class is to hide the process of creating a new data source, so the clients have no access to the real constructor of the class. Instead, clients need to use a factory method, which is implemented as a static member function of the DataSource class.

When using factory methods, it is frequently useful to hide the real implementation of the constructor. This can be done through careful use of the private modifier. The goal is to grant access to the constructor only to the class itself (and to any declared friends of the class). This is done to the standard constructor as well as to the copy constructor.

The interface to the DataSource class is presented in Listing 5-1. Both constructors and the assignment operator are declared as private. The destructor, however, needs to be accessible so that the delete keyword can be called on allocated objects. The readData member function is an interface for the main responsibility attributed to this class, and its implementation will vary according to the read data source used. The createInstance member function is a static function that creates and returns new instances of the data type, functioning as the factory method.

*Listing 5-1.* Declaration of the DataSource Class

```
//
//  DataSource.hpp

#ifndef DataSource_hpp
#define DataSource_hpp

#include <string>

class DataSource {
private:
    DataSource(const std::string &name);
    DataSource(const DataSource &p);
    DataSource &operator=(const DataSource &p);
```

```
public:
    ~DataSource();  // must be public so clients can use delete

    static DataSource *createInstance();

    void readData();

private:
    std::string m_dataName;
};
```

The implementation of the DataSource class is shown in Listing 5-2. The constructors and destructor are standard, considering the fact that the constructor is private. The interesting part of the DataSource implementation is the getInstance method, which returns a new data source. This implementation receives only one parameter that is created by the method, but consider the general case in which a list of complex or implementation-dependent objects need to be retrieved in order to call the new operator for the DataSource class.

■ **Note** At the end of getInstance, the member function returns a pointer to the newly created object. Another option is to return a smart pointer, such as std::shared_ptr, which would make it easier to manage the lifetime of the allocated object.

***Listing 5-2.*** Implementation of the DataSource Class

```
#endif /* DataSource_hpp */

//
//  DataSource.cpp

#include "DataSource.hpp"

DataSource::DataSource(const std::string &name)
: m_dataName(name)
{
}

DataSource::DataSource(const DataSource &p)
: m_dataName(p.m_dataName)
{
}

DataSource &DataSource::operator=(const DataSource &p)
{
    if (this != &p)
    {
        m_dataName = p.m_dataName;
    }
    return *this;
}
```

```
DataSource::~DataSource()
{
}

DataSource *DataSource::createInstance()
{
    std::string sourceName;
    // complex method used here to find sourceName and other construction parameters ....
    DataSource *ds = new DataSource(sourceName);
    return ds;
}

void DataSource::readData()
{
    // read data here ...
}

void useDataSource()
{
    // DataSource *source = new DataSource(""); // this will not work!
    DataSource *source = DataSource::createInstance();
    source->readData();
    // do something else with data
    delete source;
}
```

## The Singleton Pattern

One of the simplest and most used design patterns is the singleton. With this design pattern, a single object is used to represent a whole class, so that there is a central location where services managed by that class can be directed.

Unlike standard classes, a singleton class represents a single resource that cannot be replicated. Because of this, the singleton pattern restricts the ability to create new objects of a particular class, using a few techniques that will be discussed latter in this section. C++ provides all the features necessary to implement singleton patterns with high performance.

In programming, the notion of an entity that is unique across the application is frequently encountered. An example in options programming is an entity called a *clearing house*. A clearing house is an institution that provides clearing services for trades on options and derivatives. The clearing house makes sure that every trade has collateral so that counterpart risk is reduced, among other attributions. For example, if a trader sells options in a particular instrument, the clearing house will make sure that the trader has enough margin to satisfy the requirements of that particular trade.

While a clearing house provides important services in the trading industry, most applications need to connect to a single clearing house. Thus, creating a single object to represent the clearing house is an obvious implementation technique for this situation. Table 5-1 presents a few examples of objects that could be modeled using a singleton.

CHAPTER 5 ■ DESIGN PATTERNS FOR OPTIONS PROCESSING

***Table 5-1.*** *Example Objects that Can Be Implemented as a Singleton Design Pattern*

| Object | Notes |
|---|---|
| Clearing house (finance) | A single clearing house is used for all trades. |
| Root window (GUI) | Each GUI application communicates with only one root window. |
| Operating system | An object representing operating system services is unique through the application. |
| Company CEO | An object representing the CEO has only one instance. |
| Memory allocator (system services) | Each application uses a single memory allocator, which can be represented by a singleton. |

To implement a singleton in C++, the first step is to make sure that there is only one object of that class in the application. To do this, it is necessary to disallow the creation of new objects of that particular class. You can take advantage of the ability provided by C++ to make class members inaccessible to users of the class through the `private` keyword. Users then cannot use the `new` keyword to generate new objects of that particular class.

On the other hand, it is necessary to create some mechanism for clients to access an instance of the singleton class. This is usually done using a static member function that returns the single existing object, or creates a new object if necessary before returning it. Using such an access member function, clients can access the public interface of the singleton object. At the same time, they're not allowed to create or manage the lifecycle of that object.

## Clearing House Implementation in C++

A possible implementation for the clearing house class using the singleton pattern is presented in Listings 5-1 and 5-2. The class has two parts: the first part deals with the management of the singleton object. This is done through the definition of private constructor and destructors, as well as the presence of a static member function getClearingHouse, which returns a reference to the singleton instance.

The second part of the implementation deals with the responsibilities of the clearing house, represented here as the member function clearTrade. This function receives as an argument a Trade object, which is not defined here but contains all the data associated with the transaction.

Listing 5-3 shows the interface, which follows the singleton design pattern. Listing 5-4 contains the implementation of the member functions declared in the class interface, as well as the static member variable s_clearingHouse.

***Listing 5-3.*** Header File for the ClearingHouse Class, Which Implements the Singleton Design Pattern

```
//
// DesignPatterns.hpp
// CppOptions

#ifndef DesignPatterns_hpp
#define DesignPatterns_hpp

class Trade {
    // ....
};
```

CHAPTER 5 ■ DESIGN PATTERNS FOR OPTIONS PROCESSING

```
class ClearingHouse {
private:                    // these are all private because this is a singleton
    ClearingHouse();
    // the copy constructor is not implemented
    ClearingHouse(const ClearingHouse &p);
    ~ClearingHouse();
    // assignment operator is not implemented
    ClearingHouse &operator=(const ClearingHouse &p);

public:
    static ClearingHouse &getClearingHouse();

    void clearTrade(const Trade &);

private:
    static ClearingHouse *s_clearingHouse;
};

#endif /* DesignPatterns_hpp */
```

The implementation file contains the member function ClearingHouse::getClearingHouse. This function first checks the static variable s_clearingHouse to determine if it has been previously allocated. If the object doesn't exist, then the static function can create a new object, store it for further use, and return a reference.

The function useClearingHouse is an example of how the ClearingHouse class can be used. The first step is to have a variable hold a reference to the singleton object. Then, by calling the static function getClearingHouse, you can access the singleton. In this example, the singleton is used to process another trade through the member function clearTrade.

***Listing 5-4.*** Implementation File for ClearingHouse Class, Which Uses the Singleton Design Pattern

```
//
//  DesignPatterns.cpp

#include "DesignPatterns.hpp"

ClearingHouse *ClearingHouse::s_clearingHouse = nullptr;

ClearingHouse::ClearingHouse()   // private constructor, cannot be used by clients
{
}

ClearingHouse::~ClearingHouse()  // this is private and cannot be used by clients
{
}
```

```
ClearingHouse &ClearingHouse::getClearingHouse()
{
    if (!s_clearingHouse)
    {
        s_clearingHouse = new ClearingHouse();
    }
    return *s_clearingHouse;
}

void ClearingHouse::clearTrade(const Trade &t)
{
     // trade is processed here
}

void useClearingHouse()
{
    Trade trade;
    ClearingHouse &ch = ClearingHouse::getClearingHouse();
    ch.clearTrade(trade);
}
```

# The Observer Design Pattern

A frequent situation that occurs in complex systems is the occurrence of events that trigger further actions. For example, an event that happens on financial systems is the completion of an options trade. When a new trade is completed, several actions need to be performed to update the system and reflect the new positions in the ledger.

The observer design pattern is a very powerful strategy to manage event updates, based on a standard technique that gives clients the ability to listen to events and updates to a particular object and react accordingly.

There are two parts of the observer design pattern (see Figure 5-2). First there is an observer, which implements an abstract interface capable of receiving notifications. The abstract interface consists of a single member function, called notify. This member function is called by the second part of the design pattern, the Subject, when a new event occurs (the arrow between them means that the observer has a reference to the Subject object).

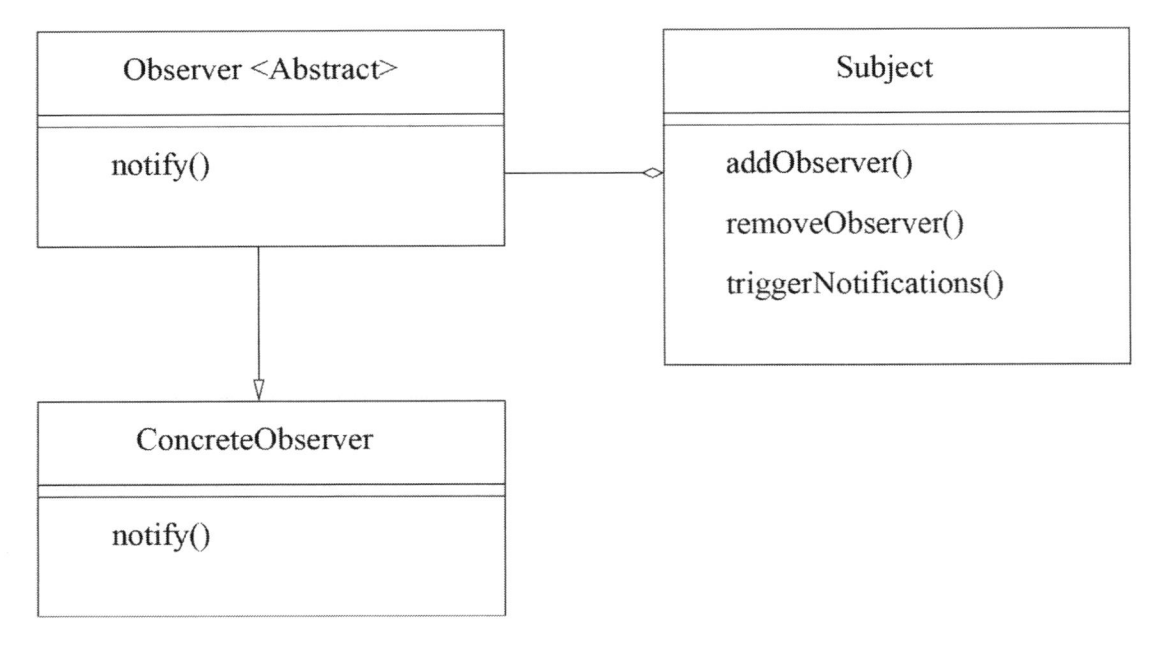

*Figure 5-2. Simplified scheme of the observer design pattern*

The Subject class has at least three member functions that enable the functionality of the observer design pattern. The first function is addObsever, which takes as a parameter a reference to an observer object. The addObserver function maintains the reference in an internal list of objects that are interested in receiving notifications.

The second member function in the subject interface is removeObserver, which simply removes a given observer from the notification list. Finally, there is a member function called triggerNotifications that is used to send the notifications to all objects that registered with the Subject class.

The observer design pattern can readily implemented in C++ using abstract classes. You can see a sample implementation in Listings 5-5 and 5-6. The first class that is considered is the Observer class. This class has the purpose of providing a simple interface for the observer. Its only non-trivial member function is notify, which is an abstract function called by the subject when a new event occurs. As a result, any class deriving from observer needs to process the notification in a user-defined way.

The interface is the following:

```
class Observer {
public:

    // constructor and destructor definitions

    virtual void notify() = 0;

};
```

■ **Note** Consider how the Observer class is independent of any implementation detail for the trading ledger system. This definition could be reused as part of a design pattern's library. Similar techniques can be used to simplify the creation of other design patterns as well.

CHAPTER 5 ■ DESIGN PATTERNS FOR OPTIONS PROCESSING

Next, it is necessary to define a class that implements the abstract observer interface. In this case, the goal is to implement a trade observer, which can be specified in the following way:

```
class TradeObserver : public Observer {
public:
    TradeObserver(TradingLedger *t);
    TradeObserver(const TradeObserver &p);
    ~TradeObserver();
    TradeObserver &operator=(const TradeObserver &p);

    void notify();
    void processNewTrade();
private:
    Trade m_trade;
    TradingLedger *m_ledger;
};
```

The constructor for this class receives as a parameter a pointer to the TradingLedger object, which will be defined later. The class provides an implementation for notifications and a member function to process new trades. These two member functions are implemented as follows.

```
void TradeObserver::notify()
{
    this->processNewTrade();
}

void TradeObserver::processNewTrade()
{
    m_trade = m_ledger->getLastTrade();
    // do trading processing here
}
```

Here, the notification implementation just calls the processNewTrade function, which stores the trade returned by the ledger object.

Finally, you can also see a definition for the TradingLedger class. The class contains the three member functions that comply with the subject interface (addObserver, removeObserver, and triggerNotifications). The class also contains two simple member functions to add and return trades, as shown in the following definitions:

```
class TradingLedger {
public:
    TradingLedger();
    TradingLedger(const TradingLedger &p);
    ~TradingLedger();
    TradingLedger &operator=(const TradingLedger &p);

    void addObserver(std::shared_ptr<Observer> observer);
    void removeObserver(std::shared_ptr<Observer> observer);
    void triggerNotifications();

    void addTrade(const Trade &t);
    const Trade &getLastTrade();
```

95

```
private:
    std::set<std::shared_ptr<Observer>> m_observers;
    Trade m_trade;
};
```

The addObserver and removeObserver functions operate with std::shared_ptr templates for the observer object. The goal is to avoid unnecessary memory issues by delegating the memory deallocation to shared pointers from the standard library. These two functions operate as an interface to the internal m_observers container.

The triggerNotification function can be implemented as follows:

```
void TradingLedger::triggerNotifications()
{
    for (auto i : m_observers)
    {
        i->notify();
    }
}
```

It simply loops through all elements stored in the m_observers set and sends a notification to these registered objects. Each such object that implements the observer interface can now respond to the event as needed.

## Complete Code

The complete example previously described can be seen in Listings 5-5 and 5-6. The first file contains only the interface for the main classes used in the system. Listing 5-6 shows the implementation of these classes, along with a sample main function that creates the ledger and two observer objects.

***Listing 5-5.*** Header File Containing Interfaces for the Observer Design Pattern

```
//
//  Observer.hpp

#ifndef Observer_hpp
#define Observer_hpp

#include <set>
#include <memory>

class Observer {
public:
    Observer();
    Observer(const Observer &p);
    ~Observer();
    Observer &operator=(const Observer &p); // not implemented
    
    virtual void notify() = 0;
    
};
```

CHAPTER 5 ■ DESIGN PATTERNS FOR OPTIONS PROCESSING

```cpp
class Trade {
    //
    // .... Implementation not shown here
};

class TradingLedger;

class TradeObserver : public Observer {
public:
    TradeObserver(TradingLedger *t);
    TradeObserver(const TradeObserver &p);
    ~TradeObserver();
    TradeObserver &operator=(const TradeObserver &p);

    void notify();
    void processNewTrade();
private:
    Trade m_trade;
    TradingLedger *m_ledger;
};

class TradingLedger {
public:
    TradingLedger();
    TradingLedger(const TradingLedger &p);
    ~TradingLedger();
    TradingLedger &operator=(const TradingLedger &p);

    void addObserver(std::shared_ptr<Observer> observer);
    void removeObserver(std::shared_ptr<Observer> observer);
    void triggerNotifications();

    void addTrade(const Trade &t);
    const Trade &getLastTrade();

private:
    std::set<std::shared_ptr<Observer>> m_observers;
    Trade m_trade;
};

#endif /* Observer_hpp */
```

***Listing 5-6.*** Implementation File with C++ Definitions for the Observer Design Pattern

```cpp
//
//  Observer.cpp

#include "Observer.hpp"

using std::shared_ptr;
```

```
typedef shared_ptr<Observer> PObserver;
typedef shared_ptr<TradeObserver> PTradeObserver;

Observer::Observer()
{
}

Observer::Observer(const Observer &p)
{
}

Observer::~Observer()
{
}

void Observer::notify()
{
}

TradeObserver::TradeObserver(TradingLedger *t)
: m_ledger(t)
{
}

TradeObserver::TradeObserver(const TradeObserver &p)
: m_trade(p.m_trade),
  m_ledger(p.m_ledger)
{
}

TradeObserver::~TradeObserver()
{
}

TradeObserver &TradeObserver::operator=(const TradeObserver &p)
{
    if (this != &p)
    {
        m_trade = p.m_trade;
        m_ledger = p.m_ledger;
    }
    return *this;
}

void TradeObserver::notify()
{
    this->processNewTrade();
}
```

```cpp
void TradeObserver::processNewTrade()
{
    m_trade = m_ledger->getLastTrade();
    // do trading processing here
}

// -- TradingLedger implementation

TradingLedger::TradingLedger()
{
}

TradingLedger::TradingLedger(const TradingLedger &p)
: m_observers(p.m_observers),
  m_trade(p.m_trade)
{
}

TradingLedger::~TradingLedger()
{
}

TradingLedger &TradingLedger::operator=(const TradingLedger &p)
{
    if (this != &p)
    {
        m_observers = p.m_observers;
        m_trade = p.m_trade;
    }
    return *this;
}

void TradingLedger::addObserver(PObserver observer)
{
    m_observers.insert(observer);
}

void TradingLedger::removeObserver(PObserver observer)
{
    if (m_observers.find(observer) != m_observers.end())
    {
        m_observers.erase(observer);
    }
}

void TradingLedger::triggerNotifications()
{
    for (auto i : m_observers)
    {
        i->notify();
    }
}
```

```cpp
void TradingLedger::addTrade(const Trade &t)
{
    m_trade = t;
    this->triggerNotifications();
}

const Trade &TradingLedger::getLastTrade()
{
    return m_trade;
}

//
// Simple test stub for the TradingLedger and TradeObserver classes.
int main()
{
    TradingLedger tl;
    PTradeObserver observer1 = PTradeObserver(new TradeObserver(&tl));
    PTradeObserver observer2 = PTradeObserver(new TradeObserver(&tl));
    tl.addObserver(observer1);
    tl.addObserver(observer2);

    // perform trading system here

    Trade aTrade;
    tl.addTrade(aTrade);

    // observers should receive a notification at this point
    return 0;
}
```

## Conclusion

Design patterns are commonly used to develop reusable code, especially when OO techniques are employed. C++ provides strong support for the creation of classes that follow designed patterns such as the ones discussed in the preceding sections.

In this chapter, you saw examples and implementation in C++ for three common design patterns. First, I presented an overview of design patterns, listing some of the patterns that are most commonly used in the implementation of algorithms for options and derivatives. Then, you learned about the factory method design pattern, which is one of the easiest and most widely used patterns of OO programming.

The singleton pattern is used when it is necessary to enforce the existence of a single instance for a particular class. You saw the example of a clearing house implementation, where the single instance must be accessible to all clients in the application.

The observer pattern is a third example of how to implement such designs in C++. You saw how this pattern can be employed to solve the problem of trading processing. Using this design pattern, it is possible to decouple the classes that receive the events from specific classes that listen to the events and perform further processing.

While object-oriented design patterns provide several elegant solutions for commonly found problems in financial programming, there are situations in which a non-OO strategy may be a better solution. In these situations, C++ promotes the use of *templates*, an implementation technique in which the compiler is allowed to generate code based on parameterized types. In the next chapter, you will see several examples in which template-based algorithms can be used to improve the performance and flexibility of algorithms for options and derivatives trading.

# CHAPTER 6

# Template-Based Techniques

C++ templates allow programmers to write generic code, which works without modification on different datatypes. Through the careful use of templates, C++ programmers can write expressive code with high performance and low overhead, without needing to rely exclusively on more computationally expensive object-oriented techniques, such as the design patterns presented in the previous chapter.

This chapter explores a few template-based programming practices that can be used to solve options- and derivatives-based financial problems. Here are some of the topics discussed in this chapter:

- *Understanding the use of templates:* You will learn about the basics of templates, including their syntax and how they can be implemented as template functions or template classes.

- *Using compile-time algorithms:* This is a quick overview of how compile-time algorithms work, with some examples such as recursive algorithms, which allow compile-time definitions that depend on themselves recursively.

- *Containers and smart pointers:* One of the most common uses of templates is to maintain containers of objects. Smart pointers are also frequently employed to simplify the code necessary for memory management.

- *Best practices:* You will learn a few best practices that will improve your template-based code.

## Introduction to Templates

A template is a way to generate parameterized code, so that different versions of the same programming definition (a class or a function) can be generated for the each given parameter. A combination of parameters can also be used when more than one parameter is required. In C++, the parameters passed to a template may be a concrete datatype (native or user-defined datatypes) or an integer number, as you will see in the following examples.

You have already seen how to use basic templates in some of the previous examples that employed standard template library containers. Such containers include vectors, maps, and sets, as provided by the C++ standard library. In this section, you will learn more about the implementation of new templates and the features they can provide to application programmers.

CHAPTER 6 ■ TEMPLATE-BASED TECHNIQUES

One of the applications of templates is to perform compile-time calculations. Performing some operations at compilation time instead of at runtime is a performance-enhancing technique that can save a lot of CPU and make your application run more smoothly. Examples of such cases include the use of integer recursive functions, conditional code that depends on particular datatypes, and container objects and smart pointers.

- *Recursive functions:* A recursive function based on integer numbers can be easily calculated using compile-time techniques. For example, some numeric algorithms depend on the use of factorials of numbers, which may be known at compilation time. Transforming a runtime computation into a compile-time transformation is an easy way to make your algorithms run faster.

- *Compile-time polymorphism:* Another example of compile-time performance enhancement is the removal of conditional code based on types. When different operations need to be performed for different types, the standard procedure in OO code is to create a hierarchy that provides a different implementation for each type involved. With templates, you can replace this type of runtime polymorphism with compilation-time polymorphism. In that case, the right template is executed based on the type that is already known at compilation time, and as a result no decision is necessary at runtime.

- *Container objects:* Container objects provide a big advantage to using templates. The STL provides several containers based on templates that streamline the task of storing objects using different memory allocation strategies. For example, `std::map`s allow programmers to map from a key type to a value type in a generic way. The use of templates also simplifies common tasks such as iterating through the elements of the container. Since templates know the type of objects stored, there is no need to use a cast or other polymorphic techniques such as is used by OO code.

- *Smart pointers:* Finally, templates also give C++ the ability to automatically manage memory using smart pointers. A smart pointer is a template that has the sole purpose of managing an object that has been passed as a pointer. The exact semantics of a smart pointer changes according to the particular template and the desired results, including for example the ability to use reference counting, or to be owned by a single client. The standard C++ library provides a small number of smart pointers, such as `std::auto_ptr`, `std::unique_ptr`, and `std::shared_ptr`, among others.

■ **Note** A possible disadvantage of templates is the possibility of duplication of code in the resulting binary application. For example, if a large template has a type parameter, the compiler needs to duplicate it for each different type with which it is used. This has the potential of creating bloated executables with several redundant compiled templates. Thankfully, modern computers have enough memory that this is not a common concern, but as application sizes grow, software developers need to consider this issue.

In the next few sections, you will see some examples of template-based techniques, and learn how these techniques can be effectively implemented in C++.

# Compilation-Time Polymorphism

One of the techniques you learned in the previous chapter is the use of polymorphism based on object-oriented features such as the `virtual` keyword. One of the advantages of templates is that they can be used to work with different types, while at the same time they avoid the need for runtime checking that is inherent to the use of polymorphic classes.

With templates, you can use compile-time polymorphism in several situations where types can be known by the compiler. This makes it possible to write code that's independent of the type used, while at the same time avoid the expense of runtime lookups.

An example that is commonly used in financial code is applying mathematical operations to different datasets. This can be done is several ways, but templates can be used to make the process efficient and transparent to the programmer. Consider the operation of normalizing a dataset. To apply such an operation to different sets, you could create a `Normalize` template, as demonstrated in the following code. First, you assume that there are two implementations available for the normalization operation, one for vectors and another for sets:

```
void array_normalize(std::vector<double> &array);

void set_normalize(std::set<double> &set);
```

The next part of this example shows the main template class, called `Normalization`. This class provides the main declaration used. In a more complete implementation, `Normalization` would contain a number of static definitions other than a single function, but that is enough to demonstrate the usefulness of the class template.

The member function `normalize` performs the work of normalization in a generic way; therefore, it must receive as argument a type that is a template parameter:

```
template <class T>
class Normalization
{
public:
    typedef T Type;
    static void normalize(T &arg);
};
```

Now, you're ready to implement as many specializations of the `normalize` function as necessary. I present two specializations here, one using a vector of doubles, and another using a set of doubles. These two implementations use the regular functions that have been declared previously, and their implementations are now shown here.

```
template <>
void Normalization<std::vector<double>>::normalize(std::vector<double> &a)
{
    array_normalize(a);
}

template <>
void Normalization<std::set<double>>::normalize(std::set<double> &a)
{
    set_normalize(a);
}
```

## CHAPTER 6 ■ TEMPLATE-BASED TECHNIQUES

■ **Note** Consider how the parameter list for the template is empty. This syntax indicates that this is a specialization of a previously defined member function.

Notice how these definitions are independent of the original class definition. This means that if you create a new type of normalization function that can be applied to a particular type, the only thing you need to do is declare a new template specialization that uses that function. Therefore, the Normalization class is essentially an open definition that can be extended by any library that decides to implement a new normalization strategy. And this can be done without any runtime overhead, since the right normalization strategy will be chosen during compilation.

Finally, I present a template function that simplifies a call to the normalization member function. This template function is called normalize and just calls the desired static member function:

```
template <class T>
void normalize(T &val)
{
    Normalization<T>::normalize(val);
}
```

Here is an example of how such a function can be called for different types. The compiler will generate optimal code by deciding which specialization of the class to use and will make the call without runtime overhead:

```
void use_normalize()
{
    std::set<double> set;
    std::vector<double> array;

    normalize(set);
    normalize(array);
}
```

## Template Functions

A template function is C++ function that can be parameterized with the use of one or more types or integral values. Using template functions you can write generic functions that work with any combination of the original parameters, expanding the domain of application for the code contained in the original function.

Consider as a first example the function returning the maximum value between the two given parameters. It is easy to write such a function for a particular datatype. For example, for integer parameters, this function can be written as:

```
int int_max(int a, int b)
{
    if (a > b)
    {
        return a;
    }
```

```
    else
    {
        return b;
    }
}
```

To create a generic version of this function, you just need to create a template function that is parameterized on the types used in the parameter list and return values.

```
template <class T>
T generic_max(T a, T b)
{
    if (a > b)
    {
        return a;
    }
    else
    {
        return b;
    }
}
```

With this template, you can not only compute the maximum of two integers, but you can also do the same for any type that supports a comparison using the > operator. This even includes non-numeric types such as strings, as you will see next.

The string case is interesting in this example, because it also involves the discussion of partial specialization. A *partial specialization* is a version of a template where one or more of the parameters have been substituted by concrete types or values. You can specialize the generic_max template function to handle zero terminated strings using a different implementation, as follows:

```
template <>
const char * generic_max(const char *a, const char *b)
{
    if (strcmp(a, b) > 0)
    {
        return a;
    }
    else
    {
        return b;
    }
}
```

This syntax indicates that this is a specialization of the previously defined generic_max function. The parameter type const char * is substituted directly in the function implementation. This function in particular uses the strcmp function from the C standard library to determine if a string is less than another.

CHAPTER 6 ▪ TEMPLATE-BASED TECHNIQUES

# Implementing Recursive Functions

One of the applications of compile-time computation through templates is the implementation of recursive functions. A *recursive* function is one in which the result of the operation for a particular value can be calculated based on the application of the same function.

The reason that it is possible to use templates for computing recursive functions is the ability of these C++ templates to take integral numbers as arguments. For example, a trivial template that prints a static value can be defined using the number as a template argument:

```
template <int N>
void printNumberPlusOne()
{
    int a = N + 1;
    std::cout << a << std::endl;
}

void usePrintTemplate()
{
    printNumberPlusOne<10>();
}
```

Here, the integer N is passed not as a function argument, but as a compile-time parameter. This means that at runtime the value of N is already known as a constant value, which makes the operation much more efficient than a normal parameter passing.

This example can be further expanded, using a recursive strategy to print N numbers at compilation time. Here is a simple version that does this recursively:

```
template<int N>
void printNumberRecursive()      // general case
{
    std::cout << N << " ";
    printNumberRecursive<N-1>();
}

template<>
void printNumberRecursive<0>()   // base case
{
    std::cout << std::endl;
}

void usePrintRecursive()
{
    printNumberRecursive<10>();
}
```

This template is implemented as a general case and a specialization (base case). The general recursion case is what should be done in most cases, which in this case is print the given template parameter N and call the same template with a small number N-1. The base case is what should happen to cause the recursion to stop. In this example, the recursion stops when the value 0 is reached, in which case the template simply prints a new line.

Taken together, these two cases for the `printNumberRecursive` template can print the numbers from N to 0 using only compilation-time expression. This means that all calculations have already been computed in the compiler, dramatically cutting down the computation effort at runtime.

You can use the same strategy to compute more complex and useful recursive functions. Table 6-1 shows a few common recursive functions that involve integer numbers and that can be easily implemented using C++ templates. Notice how these functions use their own definitions in order to compute the next value.

*Table 6-1.* Common Integer Recursive Functions

| Recursive Function | Description |
| --- | --- |
| Factorial | Calculate factorials of the form 1×2×3×...×n. |
| Fibonacci | Calculate the general recursion $F(n) = F(n - 1) + F(n - 2)$. |
| Triangular numbers | Calculate the number of items in triangular formation. |
| Binomial coefficients | Calculate the coefficients of polynomial equations of the form $(ax + b)n$. |

In a more complete example, consider the implementation of the summation of the first N integer values, for a given parameter N. You can do this with a template function that recursively calls itself. Thanks to templates, the compiler can calculate such values during compilation time. Here is an implementation of such a function:

```
template <int N>
int intSum()
{
    return N + intSum<N-1>();
}

template <>
int intSum<0>()
{
    return 0;
}

void useIntSum()
{
    std::cout << intSum<20>() << std::endl;
}
```

As before, there is a general case for most values of N and a base case that is used when the parameter is 0. The general case defines the template and its integer parameter. The base case is a template specialization, so the exact argument value needs to be provided.

The `intSum` template in the general case returns the sum of the argument that was originally passed and adds to that the value of `intSum` for N-1. Since all these calculations are based on constant values at compilation time, the result is computed using the compiler itself.

The specialization of `intSum` deals with the base case that terminates the recursion. When the argument is 0, the value 0 is returned as the value of the sum. The function `useIntSum` instantiates the template, passing the value 20 as its parameter. The result is then printed to the standard output.

# Recursive Functions and Template Classes

Recursive functions can also be implemented using template classes, instead of simple functions. This is recommended when additional information is supposed to be stored with the result of the function. A template class can also receive as a parameter an integer number, along with specializations based on that template parameter.

Consider an example template class that computes the factorial of a number. The logic of this type of computation is very similar to the functions you have seen before. However, it gives you an opportunity to see how a template class works in this situation.

```
template <long N>
class Factorial
{
public:
    enum
    {
        Argument = N
    };
    static long value();
};

template <long N>
long Factorial<N>::value()
{
    return N * Factorial<N-1>::value();
}

template <>
long Factorial<0>::value()
{
    return 1;
}

void useFactorial()
{
    Factorial<8> fact;
    std::cout << " factorial for argument " << fact.Argument << " is "
              <<  fact.value() << std::endl;
}
```

The class `Factorial` shows how a template class can store interesting values as part of the class definition. The enumeration at the beginning of the class definition contains a value called `Argument`, which stores the argument for further use. This exemplifies a feature that cannot be achieved by a simple function, since the use of a class can allow any value to be stored as an enumeration of a static variable. The way the template is expanded by the compiler is shown in Figure 6-1.

```
template <3> class Factorial {
  long value() {
    return 3 * Factorial<2>::value();
  };
};
```

```
template <2> class Factorial {
  long value() {
    return 2 * Factorial<1>::value();
  };
};
```

```
template <1> class Factorial {
  long value() {
    return 1 * Factorial<0>::value();
  };
};
```

```
template <0> class Factorial {
  long value() {
    return 1;
  };
};
```

***Figure 6-1.*** *An example of computation using template specialization. The general case of the Factorial template is instantiated with the integer 3, and new instantiations are used until the specialization for Factorial<0> is reached*

The Factorial class also contains a static function that computes the desired factorial number. As in the previous examples, the function is implemented with a general case for any integer number, and a base case, which is used when the 0 value is passed.

The useFactorial function shows how to invoke the Factorial class for a particular compile-time computation. The factorial of the value 8 is desired, so it is passed as the single argument to the template class. The next line uses the Argument enumeration value so it can retrieve the passed argument.

The value of the factorial is finally accessed using the value member function. Notice that, as usual with templates, the value function is calculated at compilation time and the result is replaced with the compiler at that particular point.

## Containers and Smart Pointers

One of the most important applications of templates in C++ is the creation of data containers. A container is a template-based object that maintains and provides access to other underlying objects or data structures. For example, a common container used in C++ is std::vector, which is a representation of sequential memory that can be accessed using a numeric index. Other more complex containers are provided in the STL and in third-party libraries that are commonly used in financial applications.

Here are some of the best-known STL containers, and the types of arguments that they expect in the standard library. A short list of available containers is displayed in Table 6-2.

■ **Note** All STL containers receive as a parameter a default Allocator type, which determines how objects are allocated, such as using the global heap or some other pre-allocated local memory. If this type is not supplied, the standard allocator for the new keyword is used when creating objects.

- std::vector<T, Allocator>: The type T passed to std::vector represents the main type of each element stored in the vector. This container guarantees that elements will be stored sequentially.

- `std::map<K, T, allocator>`: This template requires two parameters. The first parameter represents the type of the key and should be an immutable object. The second type represents the object stored for each key. Maps have variations, such as `std::unordered_map`, where entries are unordered, and `std::multimap`, where each key can have more than one associated entry.
- `std::queue<T, Allocator>`: A `std::queue` provides a first-in first-out mechanism, and the argument T is the type of elements stored in this container. This container also has a variant called `std::dequeue`, which allows elements to be removed from the front or back of the queue.
- `std::stack<T, Allocator>`: A template object that stores elements in a first-in last-out mechanism. The elements are typically allocated sequentially.

*Table 6-2. Common STL Containers and Their Parameters*

| Container Type | Description |
| --- | --- |
| `std::vector` | Container in which elements are stored in sequential mode. Each element must have the same type, as determined by the template parameter. |
| `std::map` | A container where each element is associated with a unique key. The container allows searching by keys. |
| `std::queue` | A first-in first-out container that has elements of the same type, the type being the parameter to the template. |
| `std::array` | A simple sequential group of elements that can be indexed by a number. The element type is passed as a template argument. |
| `std::list` | A linked list where each object has the same type. |
| `std:set` | A container that stores an unordered list of objects. Elements of set can be retrieved efficiently. |
| `std::stack` | A first-in last-out container where each element has the same type, as determined by the template parameter. |

The second important application of templates in the C++ standard library is in the implementation of smart pointers. A smart pointer enables you to manage the memory of objects allocated in the heap. It does this through particular strategies such as using reference counting, or restricting the access to the pointer and deleting the associated memory at the end of the current scope (as is the case with `std::auto_ptr`).

Smart pointers are possible due to the ability to generate specific code for each datatype passed as parameter. Thus, a `std::shared_ptr<OptionsContract>`, for example, can be created to manage objects of type `OptionsContract`.

Table 6-3 presents a few of the most common smart pointer templates. Some of these templates have been available as part of the standard C++ library since C++11.

CHAPTER 6 ■ TEMPLATE-BASED TECHNIQUES

*Table 6-3.* *Common Smart Pointer Templates*

| Smart Pointer | Description |
| --- | --- |
| std:auto_ptr | A smart pointer that provides automatic deallocation with single ownership semantics. |
| std:shared_ptr | A smart pointer that provides a reference counted memory management, with shared ownership semantics. |
| std::unique_ptr | A smart pointer that provides unique ownership of an object. |
| std::weak_ptr | A shared pointer that represents a weak reference to an object allocated in the heap. |

# Avoiding Lengthy Template Instantiations

C++ templates are a powerful mechanism that can be used to create generic code. With templates, it is also possible to remove undesirable code duplication, since the same code can then be applied to data of different types.

On the flip side, however, templates can also create problems due to the potential they have to slow down compilation times. Because all the code in a template is generally available to the compiler when processing translation units, it is difficult to provide separate compilation for templates. An example of a library that is victim of this behavior is boost, where typically all the functionality is included in the header files. These header files are then included each time the library is referenced in an implementation file, resulting in long build times.

Despite these shortcomings, in some situations it is possible to reduce the amount of work done by the compiler on behalf of templates. This section shows a simple technique that can be used to achieve faster template compilation speeds when desired instantiations are known ahead of time.

## Pre-Instantiating Templates

Certain templates are used in only a reduced number of cases by design. For example, consider a numeric library that creates code for different types of floating point numbers. Each class in the library can be instantiated with a particular floating point type, such as double, long double, or float. Consider for instance the following definition:

```
// file mathop.h

//
// The template class for mathematical operations
//
template <class T>
class MathOperations
{
public:
   static T squared(T value)
   {
      return value * value;
   }

  // ...
};
```

111

## CHAPTER 6 ■ TEMPLATE-BASED TECHNIQUES

This class can be used in the following way:

```
#include <mathop.h>

MathOperations<double> mathOps;

double value = 2.5;

cout << "result: " << mathOps.squared(2.5) << endl;
```

Unfortunately, because the MathOperations class is a template class, you have to include its complete definition as part of the header file, where it can be found in the compiler whenever the class is instantiated.

One possible way to reduce the size of the header file is to pre-instantiate the template for the types that you know in advance.

The first step is to remove the implementation from the header file. This is clearly possible, since you can implement class member functions outside the class declaration (whether the class is a template or not). Then, you need to add the implementation to a separate source file. Once this step is done, client code can use the template class interface, but will not be able to generate code. Therefore, for this to work, you need to instantiate the templates on the implementation file.

```
// file mathop.h

//
// The template class included by the applications
//
template <class T>
class MathOperations
{
public:
   static T squared(T value);
   // ...
};

// file mathop.cpp

//
// template member function definition
//
template <class T>
T MathOperations<T>::squared(T value)
{
   return value * value;
}

//
// function used to instantiate code for specific datatypes
//
```

112

```
void instantiateMathOps()
{
    double d = MathOperations<double>::squared(2.0);
    float f = MathOperations<float>::squared(2.0);
    int i = MathOperations<int>::squared(2);
    long l = MathOperations<long>::squared(2);
    char c = MathOperations<char>::squared(2);
}
```

In this example, I chose to instantiate five versions of the original template for numeric types. The main limitation of this technique, as I mentioned, is that your clients will not be able to generate templates for the additional types they may want to use. However, in a few situations you may really want to restrict how these templates are used, and this technique works as desired.

# Conclusion

While object-oriented design patterns provide several elegant methods for the solution of commonly found problems in financial programming, there are cases in which a non-OO strategy may be more indicated. In these situations, C++ promotes the use of templates, an implementation technique in which the compiler is allowed to generate code based on parameterized types.

In this chapter, you learned how to create new template classes and functions that use the template facilities of C++. Among other things, you saw how to create functions and classes that compute their results at compilation time. Compilation-time polymorphism, an alternative to runtime polymorphism that uses the code-generation capabilities of C++ templates, was also discussed.

The next chapter continues exploring templates in C++ with a more detailed view of the standard template library and its algorithms.

# CHAPTER 7

# STL for Derivatives Programming

Modern financial programming in C++ makes heavy use of template-based algorithms. Many of the basic algorithms related to trading options and their derivatives are implemented in terms of function and class templates. This is done due to the superior advantages of templates in performance as well as their ability to improve code reuse.

Several template-based algorithms are implemented right into the standard template library (STL), which is one of the main parts of C++ standard library. Therefore, it is important to become familiar with the concepts of algorithms in the STL, and to understand how they can be used and extended to more complex applications.

In this chapter, I discuss STL algorithms and how they can be employed in quantitative finance and other programming projects. In particular, I attempt to cover how these template-based algorithms are used in practice to solve common problems with options and other financial derivatives. After reading this chapter, you will get a better understanding of how the STL interacts with other parts of the C++ libraries, and how it imposes a certain structure on classes developed in the language.

Here are some of the concepts discussed in this chapter:

- *STL-based algorithms:* Here I present an introduction to the basic concepts of algorithms in the STL, how they interact with the container, and their basic performance characteristics.

- *Functional techniques on STL:* The STL algorithms can simplify your code with the use of a functional style of programming, whereby you can use functions as a first-class object of abstraction.

- *Working on STL containers:* STL algorithms have been developed so that they work in tandem with containers. You need to understand the usage patterns of STL algorithms and how they can efficiently employ the most common containers provided by the standard C++ library.

- *Efficient iterators:* Another way in which algorithms interact with containers is through the use of iterators. Developers can use iterators in flexible ways, thanks to the support available in the STL algorithms.

## Introduction to Algorithms in the STL

The STL offers a set of templates that can be used to solve some of the most common problems encountered in C++ programming. Among such templates you will find a list of algorithms that implement tasks such as copying, sorting, selecting, iterating, and adding elements to generic collections such as vectors, sets, maps, and their variations.

CHAPTER 7 ■ STL FOR DERIVATIVES PROGRAMMING

With STL algorithms, C++ designers created a set of template functions that manipulate generic collections. Once these algorithms have been implemented as templates, developers are free to use them for any class that satisfies the functional requirements of its container. For example, based on the STL, you can create vectors of any custom class and apply template algorithms such as sort and reverse to manipulate these objects, without having to write any additional code. Table 7-1 presents a list of algorithm types available in the STL.

*Table 7-1.* A List of Algorithm Types Available in the STL

| Algorithm Type | Description |
| --- | --- |
| Conditional testing | Performs a test of a given condition against elements of a container. Algorithms include all_of, any_of, and none_of. |
| Iteration | Performs an operation for each element of a container, such as the for_each algorithm. |
| Searching | Finds elements in a container: find, find_if, find_if_not, find_first_of, and search. |
| Counting | Returns the number of elements in a container: count and count_if. |
| Sorting | Puts the elements of the container in a defined order: sort, stable_sort, and partial_sort. |
| Partitioning | Partitions the container into two ranges according to a given property: partition, partition_copy, and partition_stable. |
| Merge | Performs the merge of two containers that have been previously sorted: merge, set_union, set_intersection, and set_difference. |
| Binary search | Implements a binary search for each STL container: lower_bound, upper_bound, and binary_search. |

The generic algorithms in the STL can be imported into a C++ application using the <algorithm> header file. Most of these algorithms are implemented directly as templates in the header file, so they can be available to any client code.

The next few sections describe a few common tasks that are implemented as STL algorithms and explain how they can be used from client code, including financial applications.

# Sorting

Sorting is a basic activity that is common to many algorithms. Therefore, it is used to employ high-performance sorting algorithms without much effort. Reusing sorting algorithms also allow programmers to avoid recreating well known algorithms and the possibility of introducing mistakes into the implementation. STL algorithms provide just what you need in order to apply sorting strategies to containers and other data structures.

The STL has a set of template algorithms that can perform sorting on many different types of containers. The right algorithm should be selected according to the desired properties of the container and the data stored in it. For this purpose, the library gives you several options corresponding to the different desired tasks and their properties. As a developer, you should become acquainted with these types of sorting algorithms. Table 7-2 lists a set of algorithms commonly available from the STL (specific implementations might add their own variants).

## CHAPTER 7 ■ STL FOR DERIVATIVES PROGRAMMING

*Table 7-2. A List of Sorting Algorithms Available in the STL*

| Sorting Algorithm | Description |
| --- | --- |
| sort | Generic sorting algorithm that can be used on most containers. This should be used in most cases. |
| stable_sort | A stable sorting procedure that maintains the relative positions of elements in the container. |
| partial_sort | An algorithm that sorts only part of a given container. |
| partial_sort_copy | An algorithm that performs partial sorting on a copy of the original container. |
| is_sorted | Returns true if the given container is already sorted. This is useful when working with an unknown container. |
| nth_element | An algorithm that sorts only one of the largest elements of a container. |

The first type of sorting template is the generic sort function. This function can be applied to a range of values that's stored in the container, given by two iterators—one for the start and another for the end of the range. As normal in the STL, the container can't beat anything that can be iterated, including arrays, vectors, maps, sets, and other container templates. This sort of function can also take as a parameter a comparison function, which is used to determine the proper order of objects in the collection.

Consider, for example, a date type. The goal is to be able to sort objects of type date, which are stored in a standard STL container. To be able to sort based on dates, however, you need to provide a comparison function for the underlying date class. In C++, this is done through the use of a functional operator that overloads the standard comparison operator. Here is a quick example:

```
class Date {
public:
    // other public methods here

    bool operator<(const Date &d);

    int year()  const { return m_year; }
    int month() const { return m_month; }
    int day()   const { return m_day; }
private:
    int m_day;
    int m_month;
    int m_year;
};

bool Date::operator<(const Date &d)
{
    if (m_year < d.m_year)
    {
        return true;
    }
    if (m_year == d.m_year and m_month < d.m_month)
    {
        return true;
    }
```

117

```
        if (m_year == d.m_year and m_month == d.m_month  and m_day < d.m_day)
        {
            return true;
        }
        return false;
}

bool operator<(const Date &a, const Date &b)
{
    return a < b;
}
```

Notice that there are two versions of the < operator. The first version is written as a member function. This is necessary so that the operator has access to the private member data of the date class. The second version of the < operator is a free function, and it is necessary when the first argument is a constant object. The implementation of the free function is directly based on the member function.

```
void sort_dates()
{
    vector<Date> dates;
    // ....   initialize the dates here

    std::sort(dates.begin(), dates.end()); //  perform comparison
}
```

The sort_dates function provides an example of using the standard sort template. In this version, the default comparison is used, which in this case is implemented by the < operator. You can, however, use a different comparison function, as shown in the following example:

```
bool year_comparison(const Date &a, const Date &b)
{
    return a.year() < b.year();
}
```

Here, the comparison is performed only using the date year fields you stored in each date object. The comparison function can be called in the following way:

```
void sort_dates()
{
    vector<Date> dates;
    // ....

    std::sort(dates.begin(), dates.end(), year_comparison);  // comparison by year
}
```

In this case, you need to provide the comparison function explicitly. The result of this sorting procedure is a sequence of dates where the elements appear in increasing order of year.

The preceding example can be used to exemplify the use of stable sorting. In a stable sort, elements that are equal with respect to the sorting strategy appear in the same relative order. This is an important feature in some sorting applications. Therefore, if you want to maintain the relative sorting position of dates within a year, you should instead use the stable_sort template function.

# Presenting Frequency Data

A simple application of sorting can be seen in the presentation of frequency data. You were given a vector of price observations, and the goal is to present this pricing data according to the frequency in which it appears. This is similar to a histogram, but the data is presented in increasing frequency, while in the histogram, the frequency is presented sequentially for each data interval.

To solve this problem, you can use STL containers and the sorting template algorithm to reorganize the results. The resulting function is called *compute frequency*. The first step is to calculate the number of bins defined by the data interval. To compute this, you'll use the variables start, end, and step size. Here is the complete implementation:

```
//
//  stl_alg.cpp
//  Sorting algorithm for price data

#include <algorithm>
#include <vector>
#include <cmath>
#include <iostream>

using std::vector;
using std::cout;
using std::endl;
using std::pair;

void compute_frequency(vector<double> &prices, double start, double end, double step)
{
    int nbins = int(std::abs(end-start)/step);

    vector<pair<int, int>> count(nbins, std::make_pair(0,0));
    for (int i=0; i<nbins; ++i)
    {
        count[i].second = i;
    }

    for (int i=0; i<prices.size(); ++i)
    {
        if (start <= prices[i]  && prices[i] <= end)
        {
            int pos = int((prices[i] - start)/step);
            count[pos].first++;
        }
    }

    std::sort(count.begin(), count.end());

    for (int i=0; i<nbins; ++i)
    {
        int k = count[i].second;
        cout << start + k * step << "-" << start + (k+1) * step << ": " << count[i].first;
    }
}
```

CHAPTER 7 ■ STL FOR DERIVATIVES PROGRAMMING

The vector count stores the frequency of each data interval. Each element of the count vector has two members—the first member is the frequency and the second member is the relative position of the interval. These two values are stored as a standard pair, and the sequence numbers are initialized in a for loop.

The next step is to store the frequency counts. This is done in a loop that iterates through the given range, adding to the frequency of each data point. Finally, after the frequencies are collected, you can sort them using the STL sort algorithm, which in this case uses the standard person operator. Following this, the frequencies are presented to standard output along with the respective ranges, which have been saved in the index variable.

```
int frequency_test()
{
    vector<double> prices = {32.3, 34, 35.6, 39.2, 38.7, 31.17, 33.14 };
    compute_frequency(prices, 31.0, 39.0, 0.1);
    return 0;
}
```

To test this code, I created a simple function that calls the computer frequency function with a few simple values. The output of the function should look like the following:

```
31-31.1: 0
31.2-31.3: 0
31.3-31.4: 0
31.4-31.5: 0
33.3-33.4: 0
// ... more data here ...
38.9-39: 0
31.1-31.2: 1
32.2-32.3: 1
33.1-33.2: 1
34-34.1: 1
35.6-35.7: 1
38.7-38.8: 1
```

Figure 7-1 shows a histogram computed from sample data processed by the function frequency test. This kind of ranking function is useful when working with financial data such as price volatility.

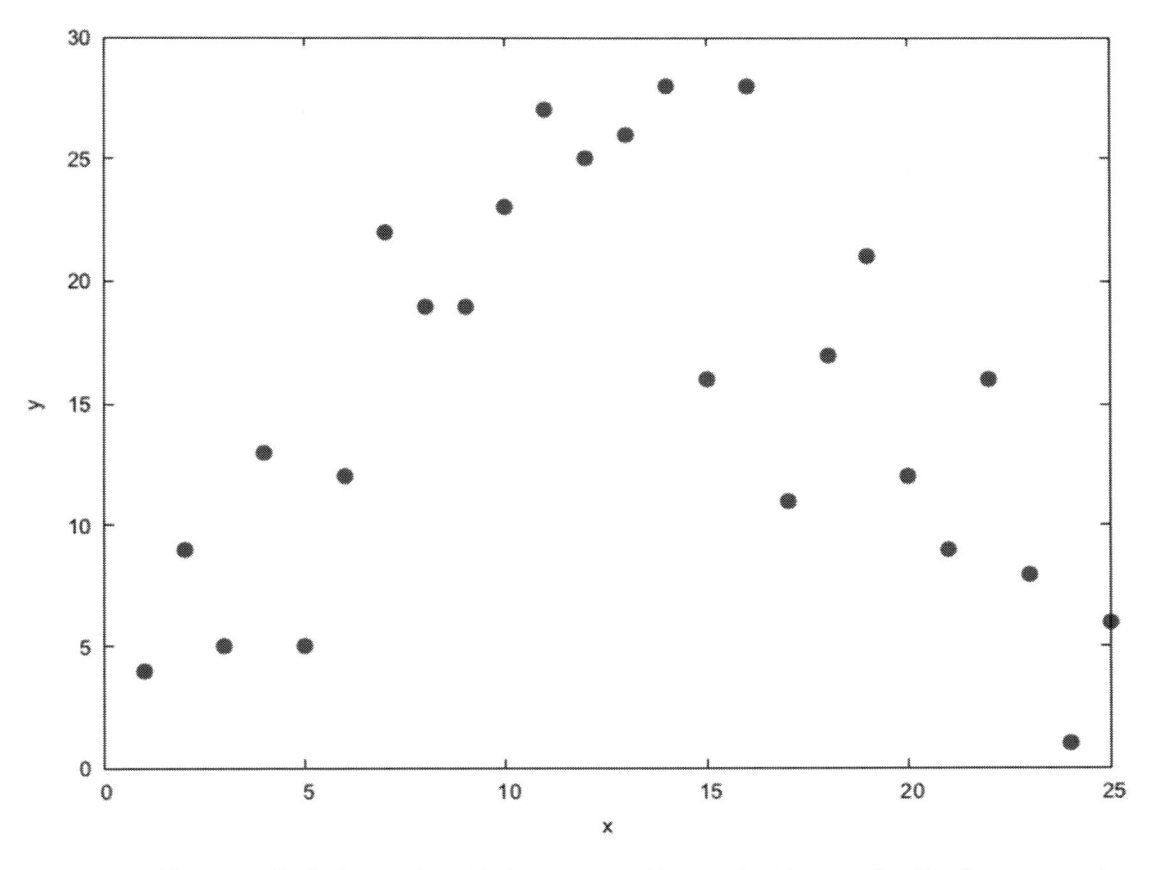

*Figure 7-1. Histogram displaying number of values computed in sample data given function frequency_test*

## Copying Container Data

Another common application of template algorithms is to copy elements from one container to another. This can be easily done using the copy template algorithm. This algorithm can perform copies between containers of different types using common conversion techniques already provided by the C++ language.

For example, it is possible to copy a container of integer numbers (int) into a second container that maintains only numbers of type double. Consider the following code:

```
void copy_int_to_double()
{
    vector<int> ivector(100, 1);
    vector<double> dvector(100);

    std::copy(ivector.begin(), ivector.end(), dvector.begin());
}
```

## CHAPTER 7 ■ STL FOR DERIVATIVES PROGRAMMING

Here, the two vectors ivector and dvector have different types. The fact that you have a template algorithm means that you don't need to write separate functions to handle every combination of types that could be presented as an argument to the copy function.

Another useful ability provided by this template is to copy elements from an existing container into the standard output. To do this, you need to wrap the standard output (or any other stream) with an std::ostream_iterator object, which allows you to iterate though an output stream. Here is an example of a simple way of displaying the contents of an STL container:

```
void print_prices()
{
    vector<double> prices(100);

    // initialize prices here

    std::copy(prices.begin(), prices.end(), std::ostream_iterator<double>(cout));
}
```

The print_prices function creates and initializes a vector of doubles. Then it passes the begin and end iterators for this vector as the first two parameters of find. Finally, the third argument wraps the standard output stream into an iterator for data of type double.

If you need to simplify the use of find (and many other similar algorithms), you could implement your own template algorithm that extracts the correct begin and end iterators. For an example of how you can do this, consider the following template function:

```
template <class T, class S >
typename T::const_iterator find(const T &a, S val) {
    return std::find (a.begin(), a.end(), val);
}
```

This template function receives two template parameters—the first is a container class and the second is a value type. The find template presented here will just call std::find and make sure that the first two arguments are the begin and end of the passed container. This code could be called similar to the previous example:

```
void find_value()
{
    vector<int> values;
    // ... initialize the vector

    vector<int>::const_iterator result = find (values, 42); // call our template
    if (result == values.end())
    {
        cout << " the value was not found " << endl;
    }
    else
    {
        cout << " the value found is " << *result << endl;
    }
}
```

122

Finally, using std::copy, it is also possible to transform a container template such as list into a different container type, such as vector. This kind of transformation allows programmers to easily convert containers of one type into another, without having to create custom code for each case. Here is an example:

```
void from_list_to_vector(const list<int> &l)
{
    vector<int> values;

    // copy contents to destination array values
    std::copy(l.begin(), l.end(), values.begin());
    // do something with the vector
}
```

In this example, the function receives a std::list of integers and copies the content stored in the list into a vector<int>. Since std::copy is a template that works with different container types, you can simply rely on the standard library to perform the desired conversions.

## Finding Elements

Finding elements in a container is another common operation that can be performed with the help of STL algorithms. The find algorithms allow programmers to search using different options. As usual, the find templates are optimized according to the specific container to which they are applied.

First you have the simple find algorithm. This algorithm takes as parameters two iterators that specify the start and end of the target data. The next parameter is a constant value that you want to find in the given container. If the value is found, the algorithm returns an iterator pointing to the desired location. If the venue is not found, the algorithm returns the second iterator, named last. Here is an example of how this works.

```
void find_value()
{
    vector<int> values;
    // ... initialize the vector

    vector<int>::iterator result = std::find(values.begin(), values.end(), 42);
    if (result == values.end())
    {
        cout << " the value was not found " << endl;
    }
    else
    {
        cout << " the value found is " << *result << endl;
    }
}
```

The find_value function is responsible for searching for a particular number inside a vector of integers. The values variable is declared as a vector container and should be initialized as desired. Next, you need to apply the find function using the beginning and end iterators returned by values. The previous example shows how to search for a constant number. The return value of this function is then stored in a vector iterator. If this variable corresponds to the end iterator, you know that the venue was not found. Otherwise, the value is printed using the contents pointed to by the iterator.

Another type of search is necessary if you use a conditional find. In this case, you should use the `find_if` template function. This function enables you to use a *predicate*, in other words, a conditional selection statement that is true only for the desired values.

Supposed for example that I try to search for a particular value inside of a container, such that the value is greater than 100. This is possible by defining a specific predicate and passing it as the last argument to the `find_if` function template. This can be done as follows:

```
bool greater_than_100(int num)
{
    return num > 100;
}

void conditional_find()
{
    vector<int> values;
    // ...  initialize the vector

    vector<int>::iterator result = std::find_if(values.begin(), values.end(), greater_than_100);
    if (result == values.end())
    {
        cout << " the value was not found " << endl;
    }
    else
    {
        cout << " the value found is " << *result << endl;
    }
}
```

First, I introduce a new predicate function called `call_greater_than_100`. This function simply returns true when the number passed as an argument is above 100. Next, you can see the function `conditional_find`. This function is similar to the previous example, but it uses the `find_if` template function instead. The first and second arguments to the `find_if` function also determine the range of values tested. The last argument is simply a pointer to the predicate function that was presented previously.

■ **Note** The last argument of `find_if` can be a function or a functional object. A *functional object* implements the function call operator, and therefore can be called using syntax similar to a call to a normal function. Such functional objects are explained in the next chapter.

## Selecting Option Data

This section shows an additional example of how STL functions can be used to speed up option data processing. This example shows a simple implementation of options, where one of the data members is the number of days until expiration.

Let the option class be defined as follows:

```
class StandardOption {
public:
    StandardOption() : m_daysToExpiration() {}
    StandardOption(int days);
    StandardOption(const StandardOption &p);
    ~StandardOption();
    StandardOption &operator=(const StandardOption &p);

    int daysToExpiration() const { return m_daysToExpiration; }

    // other function members here ...
private:
    int m_daysToExpiration;
    // other data members here ...
};

StandardOption::StandardOption(int days)
: m_daysToExpiration(days)
{
}

StandardOption::StandardOption(const StandardOption &p)
: m_daysToExpiration(p.m_daysToExpiration)
{
}

StandardOption::~StandardOption()
{
}

StandardOption &StandardOption::operator=(const StandardOption &p)
{
    if (this != &p)
    {
        m_daysToExpiration = p.m_daysToExpiration;
    }
    return *this;
}
```

This class presents a simplified version of a standard option. The number of days to expiration is stored in m_daysToExpiration and is returned by the daysToExpiration member function. You can also see a few of the standard member functions provided by the class.

The goal of this example is—given a container of StandardOptions objects—to find a set of options that are close to expiration (in this case, *closeness* is defined as a ten-day period before expiration). The first step in this process is to define a predicate function (a function returning a Boolean value), which will be called is_expiring.

```
bool is_expiring(const StandardOption &opt)
{
    return opt.daysToExpiration() < 10;
}
```

This function simply determines the number of days until expiration, and if it corresponds to the given criterion, the predicate returns true.

This predicate can be used to find all the objects of type StandardOption that satisfy the property of being close to expiration. Here is how this can be done, with the help of STL algorithms:

```
vector<StandardOption>
find_expiring_options(vector<StandardOption> &options)
{
    vector<StandardOption> result(options.size());
    std::copy_if(options.begin(), options.end(), result.begin(), is_expiring);
    if (result.size())
    {
        cout << " no expiring option was found " << endl;
    }
    return result;
}
```

First, a new vector is declared to hold the results. The final size of this vector is at most the size of the options vector. To perform the search, you can use the std::copy_if algorithm. This template algorithm copies values from the given range into the destination (result), whenever the element satisfies the given predicate function. Since you are passing a function that is true only for options close to expiration, the resulting vector will contain only near-expiration options, which are returned as the result at the end of the function.

## Conclusion

Templates allow programmers to create concise code that works on different data types. Given the advantage of templates, it is possible to create generic algorithms, which are also implemented in the core STL library. In this chapter, you learned about several template algorithms available in the C++ standard library. You also learned how to combine these algorithms to create efficient code for financial problems.

First, you saw how to use the most basic functional templates found in the STL. These include templates for tasks such as sorting coping, iterating, and accumulating values restored in a STL container.

Later, you saw examples of how to combine those functional templates into working algorithms. Template algorithms allow programmers to take full advantage of existing high-performance programming techniques coded by implementers of the C++ template library.

The use of template algorithms leads to a different style of programming, which does not to rely solely on object-oriented features. Newer versions of C++ also support functional programming. In the functional programming style, problems are solved using combinations of functions and functional objects. In these type of programs, functions are also treated as first-class objects. Treating functions this way can give you a more flexible method to organize code and solve problems. In the next chapter, you will investigate the functional style and learn how it can be used to solve financial problems occurring in options and derivatives.

# CHAPTER 8

# Functional Programming Techniques

Functional programming is a coding strategy that focuses on the direct use of functions as first-class objects. This means that in a functional program, you are allowed to create, store, and call functions, and otherwise use them as if they were just another variable of the system. Functional code also simplifies programming decisions because it avoids changing state and mutable data. This type of functional manipulation allows programs to more closely express the desired behavior of the system and is particularly suitable to some application areas. Functional programming is especially useful in the development of mathematical software and in the processing of large datasets, as is the case in the analysis options and derivatives. It also can be used with the development of multi-threaded systems, since it allows the use of lock-free code.

While the practice of functional programming was possible in previous versions of C++, such techniques have more recently been greatly improved with the adoption of the new language standard (C++11 and C++14), particularly with the introduction of lambda functions. With lambda functions, programmers can now create temporary functions in place and pass them as arguments in the call to other functions. Such features have made it easier to apply functional programing techniques to C++ applications.

In this chapter, you will learn how to use functional programming strategies to solve typical problems that occur in algorithms for trading options and derivatives. The following topics are explored in this chapter:

- *Functional objects*: A functional object allows an instance of a class to be called with the same syntax as a function, by defining the function call operator.

- *Functional templates*: The STL has support for functional programming through the use of functional templates. With them, programmers can pass functions as parameters, as well as compose functions with other functions.

- *Lambda functions*: With the introduction of C++11, a new syntax was created to represent unnamed functions, also known as *lambda functions*. You will see how to use lambdas in C++ and learn how they simplify the creation and maintenance of functional code.

- *Functional techniques for options processing:* Throughout the chapter, you will see examples of how these functional programming techniques can be effectively used to solve some problems occurring in the analysis of options and derivatives.

CHAPTER 8 ■ FUNCTIONAL PROGRAMMING TECHNIQUES

# Functional Programming Concepts

Functional programming has its roots in the analysis of mathematical algorithms, where functions are the main abstraction. Such functions are typically used to compute results based on mathematical properties of numbers. Functions can be used to express mathematical algorithms, as well as used as an effective abstraction for the creation of complex algorithms in several areas.

In particular, functional programming uses functions as building blocks to create solutions for computational problems. With this programming technique, you can call functions, as well as perform operations on these functions such as composition, partial application, currying, filtering, among others. You will see examples of this later in this chapter.

Here are a few advantages of using functional programming:

- It is possible to compose functions to achieve complex behavior from simple initial functions. Functional composition can be more easily done when functions are treated as a first-class object, instead of as an isolated element of the language.

- Functional programming doesn't depend on complex states stored inside objects. Functions are generally transparent, and they depend on arguments that are passed at each function call. In comparison, objects are complex and store a lot of context that may be hard to understand. The use of functional programming techniques favor the creation of simpler code with less state, since the state needs to be passed at each function call.

- Operations such as *memoization* can be easily performed when functions are first-class objects. With memoization, it is possible to cache the values of function calls, so that the next time a result can be immediately returned. This can be done because functions don't store any mutable state.

- No complex hierarchy of objects is necessary. Unlike OO programming, functional techniques are not based on hierarchies, and therefore require no knowledge of the internal relationships of classes. Functions are independent of each other and can be applied in any sequence.

In the next few sections, you will see examples of these functional concepts applied to C++ through different techniques. First, you will see how to use function objects for this purpose. Then, you will see how to use external libraries such as boost::lambda. Finally, you will see how to implement functional programming techniques using C++ lambda functions.

# Function Objects

The first step toward working with functional programming in C++ is to use a flexible representation for functions. One of the most common techniques for doing this is to use function objects. A function object (also known as a *functor*) is a C++ concept that allows programmers to create class instances that behave as if they were functions. The key for this concept to work is the overloading of the function call operator (represented in C++ by a pair of matching parentheses).

The function call operator can be defined as a member function in each class that needs to simulate a function call. The function call operator is called automatically from the compiler when the function call syntax is used. Consider the following example of how this process works. The OptionComparion class defines instances of a function object that compares two options (defined here using the class SimpleOption), as defined here:

```
// a simple option representation
class SimpleOption {
public:
    // other definitions here
    int daysToExpiration() const { return m_daysToExpiration; }
private:
    int m_daysToExpiration;
};
```

The first part of the code declares a class that contains options data. In this example, SimpleOption contains only the number of days to expiration. In a normal application, this class would contain a complete representation of the attributes of an option.

```
class OptionComparison {
public:
    OptionComparison(bool directionLess);
    OptionComparison(const OptionComparison &p);
    ~OptionComparison();
    OptionComparison &operator=(const OptionComparison &p);

    bool operator()(const SimpleOption &o1, const SimpleOption &o2);
private:
    bool m_directionLess;
};
```

The OptionComparison class is the main focus of this example, since it declares a datatype that can be used as a comparison function.

For the purposes here, the most important part of OptionComparison is the declaration of a function to handle the function call syntax, using operator(). In this example, the arguments passed to the function call operator are two objects of type SimpleOption that you want to compare. The following code shows the details of the implementation for OptionComparison class:

```
OptionComparison::OptionComparison(bool directionLess)
: m_directionLess(directionLess)
{
}

OptionComparison::OptionComparison(const OptionComparison &p)
: m_directionLess(p.m_directionLess)
{
}
```

```cpp
OptionComparison::~OptionComparison()
{
}

OptionComparison &OptionComparison::operator=(const OptionComparison &p)
{
    if (this != &p)
    {
        m_directionLess = p.m_directionLess;
    }
    return *this;
}

bool OptionComparison::operator()(const SimpleOption &o1, const SimpleOption &o2)
{
    bool result = false;

    // check components of opt1 and opt2. In practice this could be more complex.
    if (m_directionLess)
    {
        result = o1.daysToExpiration() < o2.daysToExpiration();
    }
    else
    {
        result = o1.daysToExpiration() > o2.daysToExpiration();
    }
    return result;
}
```

The first part of the implementation contains a few standard member functions that are required by C++. The next part of the implementation, containing operator(), shows how the comparison functionality is handled by this class. In this simple case, the class considers the m_directionLess flag to determine if a *less than* test should be used. Otherwise, the function uses a *greater than* test and returns the results.

The following function shows how to use OptionComparison:

```cpp
void test_compare()
{
    OptionComparison comparison(true);

    SimpleOption a, b;
    // ...
    // Initialize options a and b here...

    if (comparison(a, b))
    {
        std::cout << " a is less than b " << std::endl;
    }
    else
    {
        std::cout << " b is less than a " << std::endl;
    }
}
```

The first line of test_compare creates a new instance of the comparator object. Then, the code creates two SimpleOption objects and initializes it as necessary. The comparison object is then called as if it were a function, using the operator ().

The strategy displayed previously can be used to simulate functions with different signatures by creating the appropriate version of the operator (). Also, a single class can decide to implement several versions of the operator(), depending on the ways it wants to be called.

## Functional Predicates in the STL

As you saw in the previous section, objects can be used to simulate functions in C++ through the definition (or overloading) of the function call operator. This flexible mechanism can be used to create code that behaves as a function but encapsulates complex properties, as any object can do.

Based on the use of function objects, you can build a different style of programming. To facilitate the creation of functional code, the authors of the STL provide a set of basic function templates and classes that automate many common tasks. Some of these template functions are listed on Table 8-1.

*Table 8-1. List of Functional Templates Provided by the STL*

| Functional Template | Description |
| --- | --- |
| equal_to | Compares two parameters and determines equality between them. |
| greater | Compares the two given parameters and returns true if the first parameter is greater than the second. |
| greater_equal | Compares the two given parameters and returns true if the first parameter is greater than or equal to the second. |
| less | Compares the two given parameters and returns true if the first parameter is less than the second. |
| less_equal | Compares the two given parameters and returns true if the first parameter is less than or equal to the second. |
| logical_and | Receives two Boolean parameters and returns true if both parameters evaluate to true. |
| logical_or | Receives two Boolean parameters and returns true if at least one of the parameters evaluates to true. |
| logical_not | Receives a Boolean parameter and returns true if the parameter evaluates to false. |
| plus | A functional template that receives two numeric parameters and returns their sum. |
| minus | A functional template that receives two numeric parameters and returns the first minus the second. |
| negate | A functional template that receives a single numeric parameter and returns the negative of that value. |
| divides | A functional template that receives two numeric parameters and returns the value of the first parameter divided by the second. |
| bind | Receives a function or functional object as a parameter and binds the parameters to that function to constant values or to variable placeholders. |

CHAPTER 8 ■ FUNCTIONAL PROGRAMMING TECHNIQUES

The goal of the functional objects included in the STL is to provide a set of basic operations for creating new functional objects. Notice that through the combination of the given objects, it is possible to create complex functions to encode application-dependent logic. You can freely combine these functional templates to define larger expressions in a way that represents the desired functionality.

■ **Note** Be aware of the differences between using functional objects and normal C++ operations. A C++ computation specified with operators such as * and + cannot be passed as parameters to other functions, because they are immediately executed in place. Functional objects, on the other hand, form expressions that can be passed to other functions. Moreover, the process of putting these functional objects together is performed be the compiler. This ability to create complex expressions and pass them to other functional objects and templates makes these STL templates useful for the purpose of functional programming.

Consider the following examples of using these functional templates in C++. The first example shows how to use these functional templates to create a sorting predicate.

```
#include <functional>

void test_operator()
{
    using namespace std;

    vector<int> numbers = { 3, 4, 2, 1, 6 };

    sort(numbers.begin(), numbers.end(), greater<int>() );

}
```

Here, you first create a sequence of integer values and store it in the variable numbers. In this case, the code is taking advantage of the initialization syntax of C++11, which allows for the sequence type to be left unspecified, while the result is stored in a std::vector.

The next step is to call std::sort on the sequence of numbers. As you have seen before, the last argument of std::sort is a comparison function. Here, you can pass a functional object declared in functional.h, therefore freeing you from having to define a separate function.

Another simple application is to transform two sequences into a third sequence. For example, one can use the plus function to add elements from two lists:

```
void test_transform()
{
    using namespace std;

    auto list1 = { 3, 4, 2, 1, 6 };
    auto list2 = { 4, 1, 5, 3, 2 };

    vector<int> result(list1.size());
```

```
    transform(list1.begin(), list1.end(), list2.begin(), result.begin(), plus<int>() );

    // use transformed list here...

    copy(result.begin(), result.end(), std::ostream_iterator<int>(cout, ", "));

    // prints 7, 5, 7, 4, 8,
}
```

This example shows you how to take two lists and perform an arithmetic operation with its respective elements. The operation in this case is the plus functional template, which adds two values and returns the sum. The first step is to create the two sequences. You can use the auto keyword to simplify the declaration of these sequences; they will be represented as vectors of integers. A result vector is also necessary, as declared in the next line of code.

The next step is to use the std::transform function to perform a transformation from the two source sequences into the destination sequence. Each step of the transformation uses the std::plus function. The result of this process is then sent to the standard output using the std::copy template function.

You could modify this example to perform any of the arithmetical or logical operations available in the functional header file, including adding, subtracting, multiplying, and dividing. More complex operations could be performed by combining these operations.

■ **Note** In general, the transform function template is very useful when you want to perform a common operation to a list of elements. By using transform, you can reduce the number of explicit for loops in your code, making the resulting program easier to understand.

# The Bind Function

In the last section you saw that several common operations are provided in the standard library using the mechanisms of functional programming. With these templates, you can write transformations to lists of data without having to program explicit loops or use other imperative programming techniques.

However, just using the primitive operations such as subtract and divide is not enough to create complex application logic. Another thing that you can do using the techniques of functional programming is bind parameter values for a given function, so that you can have a new, modified function as a result.

Consider, for example, the std::plus<T> function provided in the functional header file. It can be used to add two numbers and can be applied to members of separate containers using the transform function template. A simple modification of this function is to have a constant number as the first parameter, so that the resulting function is in fact adding a constant value each time it is applied. Functional programming allows functions to be modified in this way, before they are applied to the required data.

The solution in the STL is provided through the std::bind function. With std::bind, you can bind a particular value to one of the arguments of a given template function. By doing this, you can create as many different functions as there are new combinations of arguments.

To use std::bind, you need to determine the function to be modified, then a sequence of values that will be bound to the function arguments. Among these bound parameters, you can also refer to the arguments supplied by the user of the function, at the time that the function is called. These arguments are called *placeholder arguments,* and named as the special variables _1, _2, _3, and so on.

CHAPTER 8 ■ FUNCTIONAL PROGRAMMING TECHNIQUES

Consider the following example of the std::bind function:

```
void use_bind()
{
    using namespace std;
    using namespace std::placeholders;

    auto list1 = { 3, 4, 2, 1, 6 };

    vector<int> result(list1.size());

    //  add 3 to each element of the list
    transform(list1.begin(), list1.end(), result.begin(), bind(plus<int>(), _1, 3));

    copy(result.begin(), result.end(), std::ostream_iterator<int>(cout, ", "));

    // prints 6, 7, 5, 4, 9,
}
```

In this example, the goal is to use a modification of the std::plus function, so that each element of the list is added to the value 3, resulting in a new vector with the results. The example is similar to what you have seen in the previous code fragment, but the bind template now modifies the plus function.

The first two lines of the example are importing std and std::placeholders namespaces. The std::placeholder namespace allows you to write the name of placeholders variables _1 or _2. Then, the original list is created and a result vector is allocated.

The transform function performs the desired changes, and bind is used to create the operation applied to each element of the list1 vector. As seen in the previous example, there are two arguments for std::plus. These arguments need to be specified in sequence. This is indicated with the second and third parameters of std::bind. The first argument is supposed to be the placeholder for the first parameter. The second argument is bound to a constant number.

The std::bind template can be used in more complex situations. For example, it can be used to find member functions for existing classes. The following example shows how bind can be used to create a variation of a member function for the SimpleOption class.

```
class SimpleOption {
public:
    // other definitions here

    double getInTheMoneyProbability(int numDays, double currentUnderlyingPrice) const;

};

auto computeInTheMoneyProblExample(const std::vector<SimpleOption> &options) ->
std::vector<double>
{
    using namespace std;
    using namespace std::placeholders;

    double currentPrice = 100.0;

    vector<double> probabilities(options.size());
```

```
    auto inTheMoneyCalc = bind(&SimpleOption::getInTheMoneyProbability, _1, 2,
    currentPrice);

    transform(options.begin(), options.end(), probabilities.begin(), inTheMoneyCalc);

    return probabilities;
}
```

This assumes that `SimpleOption` contains a member function that calculates the probability that a particular option will be in the money, given a number of days before expiration and the current underlying price. Moreover, the goal is to create function that will receive a vector of options and return the associated probabilities for the specific case of two days before expiration. The function is called `getInTheMoneyProblExample` in the previous fragment.

To do this using the STL functional algorithms, you need to find a way to express the desired condition as a functional object and pass the resulting object to `std::transform`. This can done with the help of `std::bind`. The idea is to use `std::bind` to bind the value of the first argument, which in this case is the number 2. Then, the placeholder `_1` indicates that the argument passed to the resulting function is used as the second argument to `getInTheMoneyProbability`. The bound function is then saved to a variable called `inTheMoneyCalc` and used as an argument to `transform`, applied to the `options` vector.

## Lambda Functions in C++11

As you saw in the previous sections, classes, templates, and objects can be used to represent functions and other functional objects. Unfortunately, using classes for functional programming requires you to define a function outside the current place where it is being used, thus making the process more difficult than it needs to be. Functional templates such as `std::plus` and `std::multiplies` help make this easier, but it is still not as easy as writing standard C++ code.

Other languages such as Lisp and Python have simplified this task with the concept of unnamed functions, also called *lambdas*. These unmanned functions can be passed as parameters to other functions and objects, and can be freely combined into more complex functions. This way, functional programming techniques become much easier to implement and test, when compared to languages in which functions can be created only as a static entity.

One of the big changes in C++11 was the introduction of lambda functions as a syntactical element. With the addition of lambda functions, it is now possible to create unnamed functions that can be saved as variables or passed as parameters to other functions. This considerably simplifies the task of applying functional techniques in C++ programs, as you will see in the next few examples.

A C++ lambda is a piece of C++ code that can be saved and/or passed as a parameter to other functions. With lambdas, the compiler has enough information to understand that the function will run later, probably in an environment that is independent of the current function.

The syntax of lambda functions starts with a pair of square brackets, followed by arguments and a block of code. Here is an example:

```
void use_lambda()
{
    auto fun = [](double x, double y) { return x + y; };

    double res = fun(4, 5);

    std::cout << " result is " << res << std::endl;
}
```

Here, the lambda function is introduced by [ ], followed by parameters of type double. The function simply adds the two given parameters. The compiler can deduce the result type for this lambda function. However, you can also declare the return type as part of the code, using the -> syntax:

```
auto fun = [](double x, double y) -> double { return x + y; };
```

Lambda functions can also refer to variables that have been declared outside the block of the lambda function. This makes them much more convenient than standard functions, which are independent of the surrounding variables. This process is called *lambda capture,* and it allows a lambda function to access the data stored even in a local variable, after the current function has returned.

There are two types of lambda capture:

- *Lambda capture by value:* Allows lambda functions to use the value stored in a variable that is accessible at the moment of the lambda declaration. The value can be used even after the original variable no longer exists. This is indicated by adding the name of the variable inside the square brackets that introduce the lambda function.

- *Lambda capture by reference:* This allows a lambda function to modify the variable itself, instead of just using its value. This type of capture is indicated with an & operator before the name of the desired variable.

Here is an example of both cases of lambda capture.

```
void use_lambda2()
{
    int offset = 5;

    auto fun1 = [ offset](double x, double y) { return x + y + offset; };
    auto fun2 = [&offset](double x, double y) { return x + y + offset; };

    double res = fun1(4, 5);

    std::cout << " result is " << res << std::endl;

    offset = 10;
    std::cout << " result of fun1 is " << fun1(4, 5) << std::endl;
    std::cout << " result of fun2 is " << fun2(4, 5) << std::endl;
}
```

The function named fun1 has been created with a capture to the offset variable. This capture is by value only, so it will always reflect the original value of that variable, in this case the number 5. The second lambda function fun2 captures the variable offset by reference. This means that each time fun2 is called, it will use a reference to the updated value of the offset. When the variable offset changes from 5 to 10, this will change the results produced by fun2, but will not change the results of the application of fun1, as shown in the following output:

**result is 14**
**result of fun1 is 14**
**result of fun2 is 19**

CHAPTER 8 ■ FUNCTIONAL PROGRAMMING TECHNIQUES

A lambda function can also be passed as an argument to other functions. When this happens, the compiler creates a template object of type std::function<> that stores all the information used by the lambda function. You can create new functions that receive lambdas and freely use them in your code. The compiler will automatically convert a lambda into an object during the function call. Consider the following example:

```
void use_function(std::function<int(int,int)> f)
{
    auto res = f(2,3);

    std::cout << " the function returns the value " << res << std::endl;
}
```

This function just received a std::function object and displays its result when applied to the values 2 and 3. The important part of this code is noticing that std::function defines both the return type as well as the types for each of the parameters of the given function. You can see how this information is used in the compiler with two sample lambda functions that are passed to use_function as follows:

```
void test_use_function()
{
    auto f1 = [] (int a, int b) { return a + b; };
    auto f2 = [] (int a, int b) { return a * b; };

    use_function(f1);
    use_function(f2);
}
```

When called, test_use_function will produce the following results, as expected:

**the function returns the value 5**
**the function returns the value 6**

## Complete Code

The complete code for this chapter is implemented in the Functional.hpp and Functional.cpp files. The functional techniques presented here have as dependencies only the main STL header files.

```
//
// Functional.hpp

#ifndef Functional_hpp
#define Functional_hpp

class SimpleOption {
public:
    // other definitions here
    int daysToExpiration() const { return m_daysToExpiration; }

    double getInTheMoneyProbability(int numDays, double currentUnderlyingPrice) const ;
private:
    int m_daysToExpiration;
};
```

137

CHAPTER 8 ■ FUNCTIONAL PROGRAMMING TECHNIQUES

```
class OptionComparison {
public:
    OptionComparison(bool directionLess);
    OptionComparison(const OptionComparison &p);
    ~OptionComparison();
    OptionComparison &operator=(const OptionComparison &p);

    bool operator()(const SimpleOption &o1, const SimpleOption &o2);
private:
    bool m_directionLess;
};

#endif /* Functional_hpp */

//
//  Functional.cpp

#include "Functional.hpp"

#include <iostream>
#include <vector>

#include <functional>    // for functional STL code

//
// Class SimpleOption
//

double SimpleOption::getInTheMoneyProbability(int numDays, double currentUnderlyingPrice)
const
{
    return 0; // implementation here
}

//
//   Class OptionComparison
//

OptionComparison::OptionComparison(bool directionLess)
: m_directionLess(directionLess)
{
}

OptionComparison::OptionComparison(const OptionComparison &p)
: m_directionLess(p.m_directionLess)
{
}

OptionComparison::~OptionComparison()
{
}
```

```cpp
OptionComparison &OptionComparison::operator=(const OptionComparison &p)
{
    if (this != &p)
    {
        m_directionLess = p.m_directionLess;
    }
    return *this;
}

bool OptionComparison::operator()(const SimpleOption &o1, const SimpleOption &o2)
{
    bool result = false;

    // check components of opt1 and opt2. In practice this could be more complex.
    if (m_directionLess)
    {
        result = o1.daysToExpiration() < o2.daysToExpiration();
    }
    else
    {
        result = o1.daysToExpiration() > o2.daysToExpiration();
    }
    return result;
}

void test_compare()
{
    OptionComparison comparison(true);

    SimpleOption a, b;
    // ...

    if (comparison(a, b))
    {
        std::cout << " a is less than b " << std::endl;
    }
    else
    {
        std::cout << " b is less than a " << std::endl;
    }
}

void test_operator()
{
    using namespace std;

    vector<int> numbers = { 3, 4, 2, 1, 6 };

    sort(numbers.begin(), numbers.end(), greater<int>() );

}
```

CHAPTER 8 ■ FUNCTIONAL PROGRAMMING TECHNIQUES

```cpp
void test_transform()
{
    using namespace std;

    auto list1 = { 3, 4, 2, 1, 6 };
    auto list2 = { 4, 1, 5, 3, 2 };

    vector<int> result(list1.size());

    transform(list1.begin(), list1.end(), list2.begin(), result.begin(), plus<int>() );

    // use transformed list here...

    copy(result.begin(), result.end(), std::ostream_iterator<int>(cout, ", "));

    // prints 7, 5, 7, 4, 8,
}

void use_bind()
{
    using namespace std;
    using namespace std::placeholders;

    auto list1 = { 3, 4, 2, 1, 6 };

    vector<int> result(list1.size());

    //  add 3 to each element of the list
    transform(list1.begin(), list1.end(), result.begin(),  bind(plus<int>(), _1, 3));

    copy(result.begin(), result.end(), std::ostream_iterator<int>(cout, ", "));

    // prints 6, 7, 5, 4, 9,
}

auto computeInTheMoneyProblExample(const std::vector<SimpleOption> &options) -> std::vector<double>
{
    using namespace std;
    using namespace std::placeholders;

    double currentPrice = 100.0;

    vector<double> probabilities(options.size());

    auto inTheMoneyCalc = bind(&SimpleOption::getInTheMoneyProbability, _1, 2, currentPrice);

    transform(options.begin(), options.end(), probabilities.begin(), inTheMoneyCalc);

    return probabilities;
}
```

```
void use_lambda()
{
    auto fun = [](double x, double y) -> double { return x + y; };

    double res = fun(4, 5);

    std::cout << " result is " << res << std::endl;
}

void use_lambda2()
{
    int offset = 5;

    auto fun1 = [ offset](double x, double y) -> double { return x + y + offset; };
    auto fun2 = [&offset](double x, double y) -> double { return x + y + offset; };

    double res = fun1(4, 5);

    std::cout << " result is " << res << std::endl;

    offset = 10;
    std::cout << " result of fun1 is " << fun1(4, 5) << std::endl;
    std::cout << " result of fun2 is " << fun2(4, 5) << std::endl;
}

void use_function(std::function<int(int,int)> f)
{
    auto res = f(2,3);

    std::cout << " the function returns the value " << res << std::endl;
}

void test_use_function()
{
    auto f1 = [] (int a, int b) { return a + b; };
    auto f2 = [] (int a, int b) { return a * b; };

    use_function(f1);
    use_function(f2);
}

//
// The main entry point for the test application
//
int main()
{
    test_use_function();
    return 0;
}
```

You can compile this code using any standards-compliant compiler, such as gcc (which is available for all major platforms). The following command line can be used to compile the application called Functional:

```
g++ -std=gnu++11 -o Functional Functional.cpp
```

## Conclusion

Using templates is a good way to organize your code into generic functions that work across different datatypes. However, it's only when you start to compose these functions that you start to reap the benefits of a functional programming style. Functional tools in the STL and other libraries allow programmers to use functions as first-class objects.

In this chapter, you learned a few of the techniques available for programmers who want to explore functional programming in C++. Some of these techniques include the use of functional objects, which implement the function call operator to simulate native functions. The STL provides several template functions that support the use of functional objects.

You have also seen how to create and use lambda functions, a new syntactical element introduced in C++11. With lambda functions, programmers can create unnamed functions that can be saved as variables or passed as parameters to other functions. Even more interestingly, such lambda functions can refer to variables that occur in the environment in which they were created. In this way, lambda functions reduce the need to create additional classes just for the purpose of simulating function calls.

This chapter concludes the book's overview of C++ programming techniques used on derivatives programming. In the next chapter, you will start to learn about mathematical tools that can be used to price and analyze options and other derivatives. In particular, you will learn about linear programming methods and their C++ implementations.

# CHAPTER 9

# Linear Algebra Algorithms

Linear algebra techniques are used throughout the area of financial engineering, and in particular in the analysis of options and other financial derivatives. These techniques are used for example to calculate the value of large portfolios, or to quickly price derivative instruments. This chapter contains an overview of LA algorithms and their implementation in C++.

Linear algebra algorithms consist of simple operations on sets of values arranged as vectors or matrices. There is a rich mathematical theory behind the use of vectors and matrices. Although it is out of the scope here to explain this mathematical theory, it is nonetheless important to understand how such algorithm can be implemented in C++.

It is important to understand how the traditional methods of linear algebra can be applied in C++. As a high-performance language, C++ has been used by software engineers to efficiently encode numeric algorithms, such as the ones used in linear algebra. With this goal in mind, this chapter presents a few examples that illustrate how to use some of the most common linear algebra algorithms. In this chapter, you will also learn how to integrate existing LA libraries into your code.

- *Vector operations*: Operations on vectors are some of the most common ways to explore linear algebra algorithms.

- *Implementing matrices*: A matrix is a set of numbers ordered in a two-dimensional array. Even though matrices are very common, there is no standard support for matrices in the C++ library. In this chapter, you will see how to easily create a matrix class that supports all the most common matrix operations.

- *Using linear algebra libraries*: There is a set of LA functions that have become a de-facto standard in the world of numerical computing. You will see the BLAS (Basic Linear Algebra Subroutines) and their implementations, which provide the basic functions used in most LA software (both free and commercial) available nowadays.

## Vector Operations

As you can see, LA is concerned about the mathematical properties of vector spaces. Many of the operations either produce vectors or take vectors as their arguments. Therefore, the first step to properly use LA algorithms is to have a good implementation of vectors.

On the positive side, the C++ standard library already contains an optimized container called `std::vector`, which you have used extensively in the last few chapters. On the other hand, `std::vector` doesn't implement some of the most important operations that are conventionally used in linear algebra algorithms. The first step in implementing such an algorithm is therefore to provide such operations.

CHAPTER 9 ■ LINEAR ALGEBRA ALGORITHMS

There are two kinds of mathematical operations that are needed when using vectors:

- *Operations between numbers and vectors*: Some mathematical operations involve a single number (also called a *scalar number*) and a vector as arguments. For example, you may need to multiply a vector by a scalar, or add the same number to every entry in the vector. Such operations are not available in `std::vector`, but are so common that they should be supported by any LA software package.

- *Operations between two or more vectors*: Another class or mathematical operations take two or more vectors and calculate a result based on their values. A common example is a vector product, where all members in both vectors are pairwise multiplied and finally added. Other operations like vector summation are also commonly used.

The next few examples will show how to implement some of these operations using the existing containers of the STL, such as `std::vector`.

## Scalar-to-Vector Operations

Scalar operations on vectors allow a vector to be modified by a single scalar number. The two most common scalar operations are scalar addition and scalar multiplication. You can use these operations as building blocks for more complex operations, which will be explored in the following sections.

Because the `std::vector` class is already part of the STL, the strategy used here is to create free functions (not members of a particular class) that operate on vector containers. This way, you are free to continue to use the well known functions available for `std::vector` when necessary. You can also overload these functions with other types if you feel the need to extend these definitions.

The scalar addition to vectors consists in adding the same constant number to each component of the vector. This can be implemented in the following way:

```
typedef std::vector<double> Vector;

Vector add(double num, const Vector &v)
{
    int n = (int)v.size();
    Vector result(n);
    for (int i=0; i<n; ++i)
    {
        result[i] = v[i] + num;
    }
    return result;
}
```

The first statement is a `typedef` that allows you to use the type name Vector instead of `std::vector` in this and the other examples in this chapter. Another advantage of using such a typedef in numerical algorithms like this is the possibility of changing the definition of Vector if necessary. In such a case, all the code would still compile to comply with another vector type with few or no changes.

The add function creates a new Vector with a size equal to the length of the argument vector. Then, it fills the resulting vector with the original plus the number in the first argument. Next, you can see the scalar multiplication operation:

```
Vector multiply(double num, const Vector &v)
{
    int n = (int)v.size();
```

```
    Vector result(n);
    for (int i=0; i<n; ++i)
    {
        result[i] = v[i] * num;
    }
    return result;
}
```

The multiply function is implemented similarly to add. It receives a double number and a vector. The resulting vector is created as the same size as the argument v. The resultant vector is computed element by element to comply with the definition of the scalar product operation.

These two functions create and return a new vector. This is an effective way to perform the operations, but it can be less than optimal when used in inner loops of complex algorithms. One way to speed up this process is to create a version of these functions that modify the vector in place. That is, one of the vectors is passed using a non-const reference, and it is modified to contain the result of the calculation.

Here is the scalar addition function, implemented as an in-place modifying operation:

```
void in_place_add(double num, Vector &v)
{
    int n = (int)v.size();
    for (int i=0; i<n; ++i)
    {
        v[i] += num;
    }
}
```

As you can see, this is the equivalent of the += operator, but applied to a vector and a scalar number argument. A similar implementation also works for the scalar product operation:

```
void in_place_multiply(double num, Vector &v)
{
    int n = (int)v.size();
    for (int i=0; i<n; ++i)
    {
        v[i] *= num;
    }
}
```

Last, you can take advantage of C++ operator overloading when defining these functions. With operator overloading, you can write code much more naturally, so instead of typing

```
multiply(5, add(10, a));
```

(assuming that a is a vector), you can type

```
5 * (10 * a);
```

which is much easier to understand and maintain. You can create operator versions of the previous functions using the following definitions:

```
inline Vector operator +(double num, const Vector &v)
{
    return add(num, v);
}
```

## CHAPTER 9 ■ LINEAR ALGEBRA ALGORITHMS

```
inline Vector operator *(double num, const Vector &v)
{
    return multiply(num, v);
}

inline void operator +=(double num, Vector &v)
{
    in_place_add(num, v);
}

inline void operator *=(double num, Vector &v)
{
    in_place_multiply(num, v);
}
```

Because these are inline functions, they don't add any runtime penalty to the functions that have already been defined. In fact, you can think about these definitions as shortcuts to the full definition of the vector operators, so that they are easy to type.

## Vector-to-Vector Operations

The vector-to-vector operations allow you to form mathematical expressions involving two or more vectors. The most common such operations are vector addition and vector product. They can be implemented using strategies similar to the ones used previously.

First, you will see the implementation of vector addition:

```
Vector add(const Vector &v1, const Vector &v2)
{
    int n = (int)v1.size();
    Vector result(n);
    for (int i=0; i<n; ++i)
    {
        result[i] = v1[i] + v2[i];
    }
    return result;
}
```

Here, the function allocates a resultant vector, which is populated using element-wise addition of vector entries.

```
void  in_place_add(Vector &v1, const Vector &v2)
{
    int n = (int)v1.size();
    for (int i=0; i<n; ++i)
    {
        v1[i] += v2[i];
    }
}
```

Next, you can apply the same strategy to the implementation of vector products:

```
double prod(const Vector &v1, const Vector &v2)
{
    double result = 0;
    int n = (int)v1.size();
    for (int i=0; i<n; ++i)
    {
        result += v1[i] * v2[i];
    }
    return result;
}
```

Just as you can use in-place operations for scalar-to-vector functions, you can also implement vector-to-vector operations in place, therefore saving some of the effort needed to create temporary data structures. Here are the versions of these two functions designed for in-place updates:

```
void in_place_add(Vector &v1, const Vector &v2)
{
    int n = (int)v1.size();
    for (int i=0; i<n; ++i)
    {
        v1[i] += v2[i];
    }
}

void in_place_product(Vector &v1, const Vector &v2)
{
    int n = (int)v1.size();
    for (int i=0; i<n; ++i)
    {
        v1[i] *= v2[i];
    }
}
```

Finally, you can also simplify the use of these vector operations with the help of C++ operator overloading. Instead of typing a complex set of function calls, it is much more elegant to apply the standard addition and multiplication operations whenever possible. Therefore, you can use the following inline definitions to call the given vector operations without any runtime performance penalty:

```
inline Vector operator +(const Vector &v1, const Vector &v2)
{
    return add(v1, v2);
}

inline void  operator +=(Vector &v1, const Vector &v2)
{
    in_place_add(v1, v2);
}
```

## CHAPTER 9 ■ LINEAR ALGEBRA ALGORITHMS

```
inline double operator *(const Vector &v1, const Vector &v2)
{
    return prod(v1, v2);
}

inline void  operator *=(Vector &v1, const Vector &v2)
{
    in_place_add(v1, v2);
}
```

The next operation I want to discuss is a very common function defined over a single vector. The norm of a vector can be defined as the square root of the vector product of a vector with itself. Basically, the norm of a vector is a numeric quantity that can be applied to describe the whole vector. You can very easily implement a norm in the following way:

```
double norm(const Vector &v)
{
    double result = 0;
    int n = (int)v.size();
    for (int i=0; i<n; ++i)
    {
        result += v[i] * v[i];
    }
    return std::sqrt(result);
}
```

## Matrix Implementation

In the previous section, you learned about the most basic level of linear algebra functions, dealing with single numbers and vectors. A second level of operations is defined on a two-dimensional arrangement of numbers, also known as a *matrix*. Matrices arise naturally as the result of linear algebra calculations, and they provide a convenient way to manipulate data.

Matrices are fundamental to the implementation of linear algebra algorithms that are frequently used in the analysis of options and other derivatives. Unfortunately, C++ does not support matrices directly. Programmers need to create a separate abstraction to represent a matrix or use some third-party library that contains such a datatype.

For the purpose of illustrating linear algebra and related algorithms, a Matrix class will be introduced in this section. This Matrix datatype implements some of the most common operations that are needed in a financial application. However, the Matrix class presented here doesn't include all the necessary checks that a robust implementation would require, and some of these features are left as exercise for the reader.

In particular, the Matrix class presented in this section offers the following abilities:

- Creation of square and rectangular matrices, which handle the allocation of memory for a two-dimensional container of real (floating-pointing) numbers.

- Copy constructor and assignment operator that support the basic copy operations used in C++ libraries.

- Indexing operator, so that values can be accessed with the familiar square bracket notation.

- Common linear algebra operations, such as transpose, add, and multiply, implemented as member functions.

# CHAPTER 9 ■ LINEAR ALGEBRA ALGORITHMS

The first step in defining a matrix class is to define the basic organization of the stored data. In this class, the data is stored as a sequence of rows, making maximum use of the existing vector container to help manage the data.

The header file is presented in Listing 9-1, and it includes the class declaration and a few free operators that simplify the use of the class.

*Listing 9-1.* Declarations for the Matrix Class

```
//
//  Matrix.h
//

#ifndef __FinancialSamples__Matrix__
#define __FinancialSamples__Matrix__

#include <vector>

class Matrix {
public:
    typedef std::vector<double> Row;

    Matrix(int size);
    Matrix(int size1, int size2);
    Matrix(const Matrix &s);
    ~Matrix();
    Matrix &operator=(const Matrix &s);

    void transpose();
    double trace();
    void add(const Matrix &s);
    void subtract(const Matrix &s);
    void multiply(const Matrix &s);
    void multiply(double num);

    Row & operator[](int pos);
    int numRows() const;
private:
    std::vector<Row> m_rows;
};

// free operators
//
Matrix operator+(const Matrix &s1, const Matrix &s2);
Matrix operator-(const Matrix &s1, const Matrix &s2);
Matrix operator*(const Matrix &s1, const Matrix &s2);

#endif /* defined(__FinancialSamples__Matrix__) */
```

Notice that a Row is defined as a std::vector of double numbers, using a typedef. Next, you see the usual definitions for constructors, destructors, and the assignment operator.

## CHAPTER 9 ▪ LINEAR ALGEBRA ALGORITHMS

The Matrix class contains a few common operations, implemented as member functions. Last, you see a few operator overloads, so that the class can be comfortably used along with other linear algebra types discussed previously.

The first part of the Matrix class implementation is concerned with the constructors. The class has two constructors: the first constructor creates a square matrix, that is, one that has the same number of rows and columns. This is done by instantiating each row of the matrix and adding it to the top-level m_rows vector, until the complete matrix has been allocated.

```
//
// Matrix.cpp
//

#include "Matrix.h"

#include <stdexcept>

Matrix::Matrix(int size)
{
    for (int i=0; i<size; ++i )
    {
        std::vector<double> row(size, 0);
        m_rows.push_back(row);
    }
}
```

The second way to create a matrix is to give a number of rows and a number of columns, therefore creating a rectangular matrix. The underlying algorithm is similar to the previous case:

```
Matrix::Matrix(int size, int size2)
{
    for (int i=0; i<size; ++i )
    {
        std::vector<double> row(size2, 0);
        m_rows.push_back(row);
    }
}
```

The next constructor allows you to make a copy of an existing matrix. It simply takes advantages of how vectors copy all of their contents by default. The destructor is also trivial, because of the use of std::vector to manage the data.

```
Matrix::Matrix(const Matrix &s)
: m_rows(s.m_rows)
{
}

Matrix::~Matrix()
{
}
```

The assignment operator also takes advantage of the use of an std::vector. The only thing it needs to do is copy the underlying m_rows data member.

```
Matrix &Matrix::operator=(const Matrix &s)
{
    if (this != &s)
    {
        m_rows = s.m_rows;
    }
    return *this;
}
```

The Matrix class provides an easy way to access elements, using square brackets. For this purpose, it needs to define the operator[] member function. Because an std::vector is returned, the result can also be accessed using square brackets. Therefore, if a is an object of class Matrix, users of this class can just type a[2][3] to access the forth element of the third row.

```
Matrix::Row &Matrix::operator[](int pos)
{
    return m_rows[pos];
}
```

*Transposition* is one of the most common operations in a matrix. The goal of transposition is to convert rows into columns, changing the orientation of the data stored. This class does this by creating a new set of rows, where each new row contains the elements of the corresponding column. At the end, you just need to replace the existing rows with this new set of rows. This is done using the swap member function of the underlying std::vector. This way, you don't need to worry about the details of data allocation, taking full advantage of STL data management techniques.

```
void Matrix::transpose()
{
    std::vector<Row> rows;
    for (unsigned i=0;i <m_rows[0].size(); ++i)
    {
        std::vector<double> row;
        for (unsigned j=0; j<m_rows.size(); ++j)
        {
            row[j] = m_rows[j][i];
        }
        rows.push_back(row);
    }
    m_rows.swap(rows);
}
```

Next, the Matrix class contains another very common operation called *trace*. The trace of a matrix is defined as the summation of elements in the diagonal positions of the matrix. That is, for a given matrix a, you need to sum all elements a[i][i], or in mathematical notation:

$$Trace(A) = \sum_{i=1}^{n} A_{i,i}$$

## CHAPTER 9 ■ LINEAR ALGEBRA ALGORITHMS

This function is not defined for non-square matrices.

```
double Matrix::trace()
{
    if (m_rows.size() != m_rows[0].size())
    {
        throw new std::runtime_error("invalid matrix dimensions");
    }
    double total = 0;
    for (unsigned i=0; i<m_rows.size(); ++i)
    {
        total += m_rows[i][i];
    }
    return total;
}
```

The add member function implements matrix addition. Just as with vector addition, matrix addition performs the element-wise summation of entries in the matrix. This operation is defined only when the two matrices have the same dimensions, otherwise a runtime exception is thrown.

```
void Matrix::add(const Matrix &s)
{
    if (m_rows.size() != s.m_rows.size() ||
        m_rows[0].size() != s.m_rows[0].size())
    {
        throw new std::runtime_error("invalid matrix dimensions");
    }
    for (unsigned i=0; i<m_rows.size(); ++i)
    {
        for (unsigned j=0; j<m_rows[0].size(); ++j)
        {
            m_rows[i][j] += s.m_rows[i][j];
        }
    }
}
```

The subtract operation is similar to addition. It is here just to avoid the need to multiply the whole matrix by -1 in order to do a simple subtraction.

```
void Matrix::subtract(const Matrix &s)
{
    if (m_rows.size() != s.m_rows.size() ||
        m_rows[0].size() != s.m_rows[0].size())
    {
        throw new std::runtime_error("invalid matrix dimensions");
    }
    for (unsigned i=0; i<m_rows.size(); ++i)
    {
        for (unsigned j=0; j<m_rows[0].size(); ++j)
        {
            m_rows[i][j] -= s.m_rows[i][j];
        }
    }
}
```

## CHAPTER 9 ■ LINEAR ALGEBRA ALGORITHMS

The product operation is implemented by the member function `multiply`. When you're multiplying two matrices, the resulting matrix has entries that correspond to the vector product of the i-th row and the j-th column. In mathematical notation, this is represented as:

$$(A \times B)_{i,j} = \sum_{k=1}^{n} A_{ik} B_{kj}$$

The `multiply` member function updates the matrix in place; therefore, it just needs to create a new set of rows and swap the results at the end of the function.

```
void Matrix::multiply(const Matrix &s)
{
    if (m_rows[0].size() != s.m_rows.size())
    {
        throw new std::runtime_error("invalid matrix dimensions");
    }
    std::vector<Row> rows;
    for (unsigned i=0; i<m_rows.size(); ++i)
    {
        std::vector<double> row;
        for (unsigned j=0; j<s.m_rows.size(); ++j)
        {
            double Mij = 0;
            for (unsigned k=0; k<m_rows[0].size(); ++k)
            {
                Mij += m_rows[i][k] * s.m_rows[k][j];
            }
            row.push_back(Mij);
        }
        rows.push_back(row);
    }
    m_rows.swap(rows);
}
```

The `Matrix` class also defines a `multiply` member function that performs multiplication by a scalar number. This is analogous to the scalar multiplication of vectors and multiplies each element of the matrix by the same number.

```
void Matrix::multiply(double num)
{
    for (unsigned i=0; i<m_rows.size(); ++i)
    {
        for (unsigned j=0; j<m_rows[0].size(); ++j)
        {
            m_rows[i][j] *= num;
        }
    }
}
```

## CHAPTER 9 ■ LINEAR ALGEBRA ALGORITHMS

The numRows member function just returns the number of rows in the matrix.

```
int Matrix::numRows() const
{
    return (int)m_rows.size();
}
```

Finally, three operations are defined that simplify the use of the class. These operators use the in-place implementations you have saw previously, and they allow the use of convenient expressions involving matrices. These operators just give you an idea of how this works in practice; you can extend these definitions to include other common operators, such as /, +=, and *=, for example.

```
Matrix operator+(const Matrix &s1, const Matrix &s2)
{
    Matrix s(s1);
    s.subtract(s2);
    return s;
}

Matrix operator-(const Matrix &s1, const Matrix &s2)
{
    Matrix s(s1);
    s.subtract(s2);
    return s;
}

Matrix operator*(const Matrix &s1, const Matrix &s2)
{
    Matrix s(s1);
    s.multiply(s2);
    return s;
}
```

## Using the uBLAS Library

In the previous sections, you saw simple implementations of linear algebra concepts in C++. While they are useful for the examples provided in this book, sometimes you will be required to create high-performance implementations of complex numerical algorithms involving vectors and matrices. In such cases, it is useful to use well-tested and optimized libraries that provide linear algebra related code.

The most used library for linear algebra algorithms is the LAPACK. Originally written in Fortran, LAPACK (linear algebra package) aims at providing high performing and well-tested algorithms for basic operations involving vectors and matrices.

One interesting aspect of LAPACK is that it relies on another library called BLAS (Basic Linear Algebra Subprograms) to implement basic vector and matrix routines. The result is that BLAS became a standard for implementing vector and matrix routines. Several versions of BLAS have been released, providing optimized performance for specific architectures. BLAS has also versions targeting C and C++ that are used in many commercial products and other applications that need extensive support for numerical algorithms.

BLAS defines three levels of routines for support of linear algebra algorithms:

- BLAS Level 1 supports only vector-to-scalar and vector-to-vector operations. It is the most basic level of support, upon which other levels may be built.

## CHAPTER 9 ■ LINEAR ALGEBRA ALGORITHMS

- BLAS Level 2 offers optimized routines for vector-to-matrix calculations.
- BLAS Level 3 expands the previous levels to support matrix-to-matrix calculations, including operations such as matrix multiplication.

There are several implementations of BLAS, both in Fortran as well as in C++. Boost uBLAS is an implementation that is free and mostly compatible with the original BLAS library. It contains the same three support levels listed previously.

For an example of how to use uBLAS, assume that you want to access a fast implementation of the premultiply operations. That is, given a vector and a matrix, you want to write an algorithm that multiplies the vector by the matrix, giving a vector as a result.

To solve this problem, you can import the uBLAS libraries and create a function that receives two arguments: a vector and a matrix object. Here is a possible implementation for this function:

```cpp
#include "Matrix.h"

#include <boost/numeric/ublas/matrix.hpp>
#include <boost/numeric/ublas/io.hpp>
#include <boost/numeric/ublas/lu.hpp>

namespace ublas = boost::numeric::ublas;

std::vector<double> preMultiply(const std::vector<double> &v, Matrix &m)
{
    using namespace ublas;
    ublas::vector<double> vec;
    std::copy(v.begin(), v.end(), vec.end());

    int d1 = m.numRows();
    int d2 = (int)m[0].size();
    ublas::matrix<double> M(d1, d2);

    for (int i = 0; i < d1; ++i)
    {
        for (int j = 0; j < d2; ++j)
        {
            M(i,j) = m[i][j];
        }
    }

    vector<double> pv = prod(vec, M);

    std::vector<double> result;
    std::copy(pv.begin(), pv.end(), result.end());
    return result;
}
```

The first step is to include the header files for the boost numeric libraries. (You also need to make sure that the program will link to the necessary libraries; check your boost documentation for details.) Then, a function called preMultiply is defined, receiving a vector and a matrix as its parameters.

## CHAPTER 9 ■ LINEAR ALGEBRA ALGORITHMS

One of the first things this function needs to do is convert the parameters into the types required by the uBLAS library. In particular, uBLAS provides the vector<double> and matrix<double> types. You need to convert your data to these types before calling any uBLAS functions.

Once the data has been prepared, you may call the prod function from uBLAS, which knows how to calculate the product of a vector and a matrix. The result is then saved into an std::vector container and returned to the caller.

## Complete Code

This section contains the complete code for the vector operations. These functions may be used as the basis for a complete LA package, which is a common requirement in the analysis of options and derivatives.

The code is spread over two source files—LAVectors.hpp is the header file and LAVectors.cpp is the implementation file—which you'll find in Listings 9-2 and 9-3.

*Listing 9-2.* Header File LAVectors.hpp

```
//
//  LAVectors.hpp

#ifndef LAVectors_hpp
#define LAVectors_hpp

#include <vector>

typedef std::vector<double> Vector;

// scalar by vector operations

Vector add(double num, const Vector &v);
Vector multiply(double num, const Vector &v);

void in_place_add(double num, Vector &v);
void in_place_multiply(double num, Vector &v);

inline Vector operator +(double num, const Vector &v)
{
    return add(num, v);
}

inline Vector operator *(double num, const Vector &v)
{
    return multiply(num, v);
}

inline void operator +=(double num, Vector &v)
{
    in_place_add(num, v);
}
```

156

## CHAPTER 9 ■ LINEAR ALGEBRA ALGORITHMS

```cpp
inline void operator *=(double num, Vector &v)
{
    in_place_multiply(num, v);
}

// vector to vector operations

Vector add(const Vector &v1, const Vector &v2);
void   in_place_add(Vector &v1, const Vector &v2);

double product(const Vector &v1, const Vector &v2);
void   in_place_product(Vector &v1, const Vector &v2);

inline Vector operator +(const Vector &v1, const Vector &v2)
{
    return add(v1, v2);
}

inline void  operator +=(Vector &v1, const Vector &v2)
{
    in_place_add(v1, v2);
}

inline double operator *(const Vector &v1, const Vector &v2)
{
    return product(v1, v2);
}

inline void  operator *=(Vector &v1, const Vector &v2)
{
    in_place_add(v1, v2);
}

double norm(const Vector &v);

#include <stdio.h>

#endif /* LAVectors_hpp */
```

***Listing 9-3.*** Implementation File LAVectors.cpp

```cpp
//
//  LAVectors.cpp

#include "LAVectors.hpp"

#include <cmath>
```

```cpp
//
// adds a scalar number to a vector "v"
//
Vector add(double num, const Vector &v)
{
    int n = (int)v.size();
    Vector result(n);
    for (int i=0; i<n; ++i)
    {
        result[i] = v[i] + num;
    }
    return result;
}

//
// pre-multiply a number "num" by the given vector "v"
//
Vector multiply(double num, const Vector &v)
{
    int n = (int)v.size();
    Vector result(n);
    for (int i=0; i<n; ++i)
    {
        result[i] = v[i] * num;
    }
    return result;
}

//
// perform vector addition in place (modifying the given vector)
//
void in_place_add(double num, Vector &v)
{
    int n = (int)v.size();
    for (int i=0; i<n; ++i)
    {
        v[i] += num;
    }
}

//
// perform matrix multiplication in place (modifying the given vector)
//
void in_place_multiply(double num, Vector &v)
{
    int n = (int)v.size();
    for (int i=0; i<n; ++i)
    {
        v[i] *= num;
    }
}
```

```
//
// perform vector addition of two vectors  (v1 and v2)
//
Vector add(const Vector &v1, const Vector &v2)
{
    int n = (int)v1.size();
    Vector result(n);
    for (int i=0; i<n; ++i)
    {
        result[i] = v1[i] + v2[i];
    }
    return result;
}

//
// performs the vector product of vectors v1 and v2
//
double product(const Vector &v1, const Vector &v2)
{
    double result = 0;
    int n = (int)v1.size();
    for (int i=0; i<n; ++i)
    {
        result += v1[i] * v2[i];
    }
    return result;
}

//
// in place addition of vectors v1 and v2
//
void  in_place_add(Vector &v1, const Vector &v2)
{
    int n = (int)v1.size();
    for (int i=0; i<n; ++i)
    {
        v1[i] += v2[i];
    }
}

//
// in place product of vectors v1 and v2
//
void  in_place_product(Vector &v1, const Vector &v2)
{
    int n = (int)v1.size();
    for (int i=0; i<n; ++i)
    {
        v1[i] *= v2[i];
    }
}
```

```
//
// computes the norm of a vector
//
double norm(const Vector &v)
{
    double result = 0;
    int n = (int)v.size();
    for (int i=0; i<n; ++i)
    {
        result += v[i] * v[i];
    }
    return std::sqrt(result);
}
```

## Conclusion

In this chapter, you learned about linear algebra algorithms that are commonly used in the development of software for the analysis of options and other derivatives. Linear algebra provides many of the techniques that are applied to important problems such as option pricing and the numerical approximation of certain derivatives occurring in finance.

First, you learned about the basic algorithms that involve a vector and a scalar number. These operations can be implemented in C++ using functions that are applied to standard vectors, as you saw in the given examples.

Next, you learned how to implement a useful matrix datatype. Matrices are not directly provided by the STL, but you can take advantage of existing support by vectors as a building block for matrix representations. You also learned about the basic operations that can be performed over matrix objects.

Finally, I discussed linear algebra libraries that provide some of the functionality discussed in the previous sections. In particular, BLAS has been created and improved by some of the greatest specialists in the implementation of numeric algorithms. The BLAS library is organized into different levels of support for linear algebra algorithms. You saw an example of how to take advantage of this highly optimized library to improve the performance of your own LA code.

In the next chapter, you will learn about another building block for financial derivatives: numeric algorithms used to solve mathematical equations. This type of algorithms is at the core of many techniques used in the pricing of options and more exotic derivatives, as you will see in the next few chapters.

# CHAPTER 10

# Algorithms for Numerical Analysis

Equation solving is one of the main building blocks for financial algorithms used in the analysis of options and financial derivatives. This happens because of the nature of options pricing, which is based on the Black-Scholes pricing model. Many of the techniques that involve options pricing require the efficient solution of equations and other mathematical formulations.

Given the importance of mathematical techniques in the pricing of such derivatives, it is important to be able to calculate the solution for particular mathematical models. Although this is a vast area of numerical programming, I will present a few illustrations of numerical algorithms that can be used as a starting point for developing your own C++ code.

In this chapter, you will see programming examples for a few fundamental algorithms in numerical programming. In particular, you will learn techniques to calculate equation roots and integrate functions in C++, with a discussion of how they work and how they are applied. The chapter also discusses numerical error and stability issues that present a challenge for developers in the area of quantitative financial programming.

- *Mathematical function representation*: I initially discuss a representation for mathematical functions that can be used as the starting point for algorithms that manipulate these mathematical abstractions.

- *Root finding algorithms*: One of the most common types of numerical algorithms, root finding techniques are used to find one or more *roots* of an equation, which are the points where the equations have zero value.

- *Integration algorithms*: Another common type of numerical algorithms, integration techniques are used to calculate the numerical value of an integral (which can also be described as the area under a function for single dimensional equations).

- *Numerical examples in C++*: This chapter also includes C++ code that implements many of these concepts, with concrete examples of how to code these algorithms.

## Representing Mathematical Functions

The first step in this short overview of numerical algorithms is to find a reasonable way to represent mathematical functions in C++. As you saw in the previous chapter, functions can be easily represented in C++ using functional objects, which declare a function call operator as one of its member functions. Using this strategy, it is possible to convert a class instance into a callable object, with semantics similar to native functions.

A similar strategy can be used to represent mathematical functions. The main difference between generic C++ and mathematical functions is that the latter operate only over numeric domains, more commonly using float or double values.

In the following example, a new MathFunction class is declared using this strategy. The declaration of MathFunction as an abstract interface allows programmers to extend this definition as necessary to represent concrete functions, as you will see next.

CHAPTER 10 ■ ALGORITHMS FOR NUMERICAL ANALYSIS

The abstract class can be defined as presented in Listing 10-1.

***Listing 10-1.*** Definition for the Abstract Class MathFunction

```
#include <iostream>
#include <vector>

using std::cout;
using std::endl;

class MathFunction {
public:

    virtual ~MathFunction() {}
    virtual double operator()(double x) = 0;
private:
    // this just an interface
};
```

■ **Note** Because MathFunction is a polymorphic base class, it needs to define its own virtual destructor. This is necessary because clients will receive pointers or references to the base class. Without a virtual destructor, the compiler cannot determine the right destructor to be called, and as a result such objects will not be properly cleaned up.

The great thing about using this type of interface class is that once you have a class like MathFunction, you can start writing code that uses it directly. Your code is insulated from any worries about the exact representation of objects. For example, consider a useful class called PolynomialFunction, which implements the interface described by MathFunction:

```
//
//  Polynomial has the form   c_1 x^n + c_2 x^n-1 +  ....  + c_n-1 x^1 + c_n
//
class PolynomialFunction : public MathFunction {
public:
    PolynomialFunction(const std::vector<double> &coef);
    PolynomialFunction(const PolynomialFunction &p);
    virtual ~PolynomialFunction();
    virtual PolynomialFunction &operator=(const PolynomialFunction &p);

    virtual double operator()(double x) override;
private:
    std::vector<double> m_coefficients;
};
```

The PolynomialFunction class derives from MathFunction so that it can implement the same interface. However, it is only usable to represent polynomial functions, that is, functions that are determined by a polynomial of the form

$$f(x) = c_1 x^n + c_2 x^{n-1} + \cdots + c_{n-1} x^1 + c_n$$

The polynomial is determined using the coefficients passed as vectors to the constructor of PolynomialFunction. The constructors are responsible for updating the m_coefficients data member using this information.

```
PolynomialFunction::PolynomialFunction(const std::vector<double> &coef)
: m_coefficients(coef)
{
}

PolynomialFunction::PolynomialFunction(const PolynomialFunction &p)
: m_coefficients(p.m_coefficients)
{
}

PolynomialFunction::~PolynomialFunction()
{
}

PolynomialFunction &PolynomialFunction::operator=(const PolynomialFunction &p)
{
    if (this != &p)
    {
        m_coefficients = p.m_coefficients;
    }
    return *this;
}
```

## Using Horner's Method

The main part of the PolynomialFunction class is the implementation for the method call operator. Since this class represents a polynomial, this operator needs to receive a real number x and evaluate the function at that particular point. This is done using the so-called Horner's method.

Horner's method is just a quick way to evaluate a polynomial, so that you don't need to explicitly evaluate the terms $x^i$, for i from 1 to n. This can be done using a loop, where at each step you add a coefficient and multiply the result by x. A simple implementation of this idea can be done as follows:

```
double PolynomialFunction::operator()(double x)
{
    int n = (int)m_coefficients.size();
    double y = 0;
    int i;
    for (i=0; i<n-1; ++i)
    {
        y += m_coefficients[i];
        y *= x;
    }
    if (i < n) {
        y += m_coefficients[i];
    }
    return y;
}
```

CHAPTER 10 ■ ALGORITHMS FOR NUMERICAL ANALYSIS

To test these classes, you create a sample function that evaluates a polynomial function in a particular range. The function tested here is simply $x^2$ in the real range of –2 to 2. The function also prints the results so that you can visualize the data.

```
int test_poly_function()
{
    PolynomialFunction f( { 1, 0, 0 } );

    double begin = -2, end = 2;
    double step = (end - begin) / 100.0;
    for (int i=0; i<100; ++i)
    {
        cout << begin + step * i << ", ";
    }
    cout << endl;
    for (int i=0; i<100; ++i)
    {
        cout << f( begin + step * i) << ", ";
    }

    return 0;
}
```

I ran this function and plotted the results as a graph of the function. Figure 10-1 shows the output of the plot.

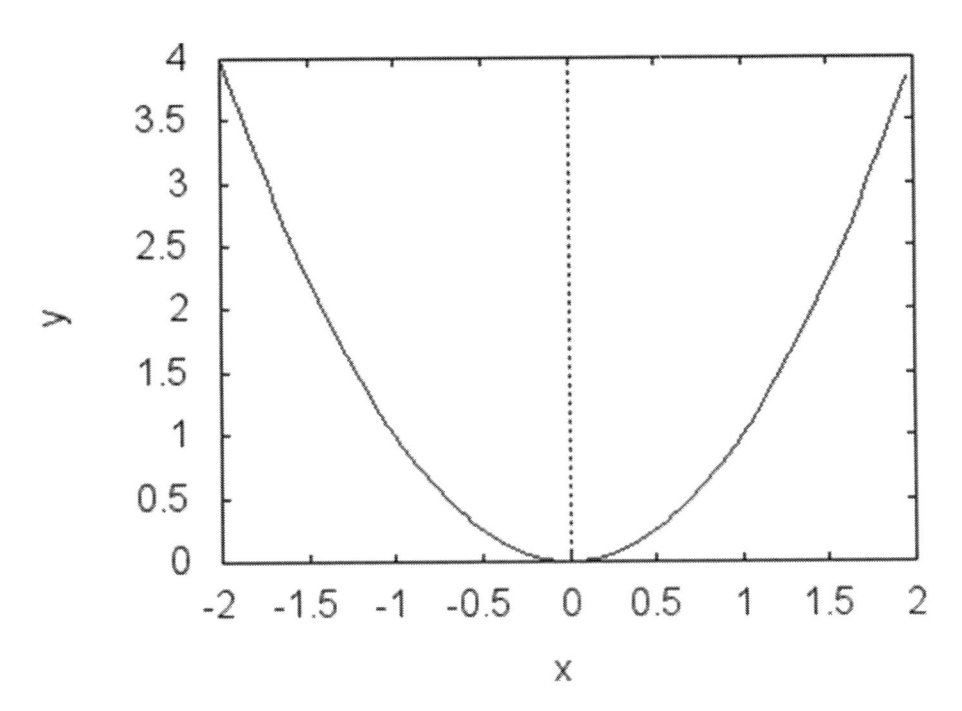

*Figure 10-1. Plot of results printed by the test_poly_function function*

# Finding Roots of Equations

Once you have a good representation for mathematical functions, it becomes possible to solve a few numerical problems. The first one I discuss is this section is finding the roots of an equation, a common problem that occurs as part of several numerical algorithms. Finding roots of an equation consists of determining one or more points in a numerical domain (usually the real numbers) where the equation has a value of zero.

This problem has a long history in mathematics, and for some types of equations it is possible to calculate their roots exactly. For example, you can find such roots for polynomials in general. For other equations, however, this problem can be too complicated to solve using analytical methods, which leads to the need for an algorithm capable of generating approximate solutions to such equations.

A number of numerical algorithms have been proposed in the mathematical literature to find the roots of equations. In this section, you see how to do this using Newton's method, which is one of the most common algorithms for this problem, and learn how it can be implemented in C++.

## Newton's Method

Newton's method is based on the use of the derivative as an approximation to the function on a particular neighborhood. To understand how this method works, notice that the derivative of a function at a particular point is known to be the slope of a line segment that is tangent to the function.

Using this property, it is very easy to improve the approximation to the equation root with a new point that is determined by the tangent. Newton's method will essentially iterate through this process, until the difference between successive approximations is very small.

This method can be readily implemented in C++ using the tools that you already have. The first part consists of creating a class that encapsulates the necessary data for the approximation procedure. Here is the definition for the NewtonMethod class:

```
#include "MathFunction.hpp"

//
// a Newton method implementation.
//
class NewtonMethod {
public:
    // Takes as parameter the function and its derivatives
    //
    NewtonMethod(MathFunction &f, MathFunction &derivative);
    NewtonMethod(MathFunction &f, MathFunction &derivative, double error);
    NewtonMethod(const NewtonMethod &p);
    virtual ~NewtonMethod();
    NewtonMethod &operator=(const NewtonMethod &p);

    double getFunctionRoot(double initialValue);
private:
    MathFunction &m_f;
    MathFunction &m_derivative;
    double m_error;
};
```

The NewtonMethod class contains the commonly used member functions and in addition it provides a function called getFunctionRoot, which receives as a parameter an initial value (a first guess that will work as a starting point).

CHAPTER 10 ▪ ALGORITHMS FOR NUMERICAL ANALYSIS

The class stores as its data a reference to the function for which you want to find roots, and another reference to its derivative. Although it is technically possible to find the derivative for most functions, the techniques to do this in a generic way are beyond the capabilities of this class, so you need to receive the derivative as a constructor parameter and store it.

```
#include <iostream>
#include <cmath>

using std::endl;
using std::cout;

namespace {
    const double DEFAULT_ERROR = 0.0001;
}

NewtonMethod::NewtonMethod(MathFunction &f, MathFunction &derivative)
: m_f(f),
m_derivative(derivative),
m_error(DEFAULT_ERROR)
{
}

NewtonMethod::NewtonMethod(MathFunction &f, MathFunction &derivative, double error)
: m_f(f),
m_derivative(derivative),
m_error(error)
{
}

NewtonMethod::NewtonMethod(const NewtonMethod &p)
: m_f(p.m_f),
m_derivative(p.m_derivative),
m_error(p.m_error)
{
}

NewtonMethod::~NewtonMethod()
{
}

NewtonMethod &NewtonMethod::operator=(const NewtonMethod &p)
{
    if (this != &p)
    {
        m_f = p.m_f;
        m_derivative = p.m_derivative;
        m_error = p.m_error;
    }
    return *this;
}
```

166

These member functions are necessary just to maintain the state of NewtonMethod objects. The m_f member stores the function that needs to be solved. The m_derivative member stores a reference to the derivative of the main function. You can also tweak the expected error of the solutions found by this class using the m_error member function. If the error is not supplied, this class uses the value stored in the DEFAULT_ERROR error constant.

Next, you're ready for the implementation of Newton's method using the given infrastructure. The getFunctionRoot function provides the necessary code for finding the root of the equation. This member function is essentially a loop in which at each step a new approximation for the function root is provided. The loop ends when the absolute difference between the two last approximations is at least equal to the acceptable error:

```
double NewtonMethod::getFunctionRoot(double x0)
{
    double x1 = x0;
    do
    {
        x0 = x1;
        cout << " x0 is " << x0 << endl;   // this line just for demonstration
        double d = m_derivative(x0);
        double y = m_f(x0);
        x1 = x0 - y / d;
    }
    while (std::abs(x0 - x1) > m_error);
    return x1;
}
```

Inside the main loop, the steps are:

1. Find the value at the derivative at the current estimate point using the m_derivative member.

2. Find the value of the function itself at the current estimate, using the m_f member.

3. The derivative gives the slope d of the tangent, which can now be used to calculate another estimate point starting from the previous estimate. The equation for this new estimate is given by

$$x_1 = x_0 - \frac{f(x_0)}{f'(x_0)}$$

where $x_0$ is the previous estimate and $x_1$ is the new estimate.

You can use a few sample functions to test the accuracy of this method. I created a SampleFunction class for this purpose. This class inherits publicly the MathFunction interface and can be used to compute the function $f(x) = (x-1)^3$, which has 1 as a root solution.

```
class SampleFunction : public MathFunction {
public:
    virtual ~SampleFunction();
    virtual double operator()(double value);
}.
```

```
SampleFunction::~SampleFunction()
{
}

double SampleFunction::operator ()(double x)
{
    return (x-1)*(x-1)*(x-1);
}
```

To use this class with NewtonMethod, you also need to supply its derivative. I have implemented the Derivative class, which again is derived from MathFunction. Simple math shows you that the derivative is given by $f'(x) = 3(x-1)^2$.

```
class Derivative : public MathFunction {
public:
    virtual ~Derivative();
    virtual double operator()(double value);
};

// represents the derivative of the sample function
Derivative::~Derivative()
{
}

double Derivative::operator ()(double x)
{
    return 3*(x-1)*(x-1);
}
```

With these two classes, you can create a simple main function that puts them together and finds the root of the desired function. This code instantiates both SampleFunction and Derivative objects and creates an object of the NewtonMethod class. Finally, the code prints the value for a given initial estimate of 100.

```
int main()
{
    SampleFunction f;
    Derivative df;
    NewtonMethod nm(f, df);
    cout << " the root of the function is " << nm.getFunctionRoot(100) << endl;
    return 0;
}
```

Running this function gives as a result a set of points, each one closer to the desired equation root. You can view the sequence of results in Table 10-1.

CHAPTER 10 ■ ALGORITHMS FOR NUMERICAL ANALYSIS

*Table 10-1.* Sequence of Values Found by Newton's Method Applied to Function $(x-1)^3$ and with Initial Guess of 100

| Iteration | Estimate | Difference |
|---|---|---|
| 1 | 100 | |
| 2 | 67 | 33.00000 |
| 3 | 45 | 22.00000 |
| 4 | 30.3333 | 14.66670 |
| 5 | 20.5556 | 9.77770 |
| 6 | 14.037 | 6.51860 |
| 7 | 9.69136 | 4.34564 |
| 8 | 6.79424 | 2.89712 |
| 9 | 4.86283 | 1.93141 |
| 10 | 3.57522 | 1.28761 |
| 11 | 2.71681 | 0.85841 |
| 12 | 2.14454 | 0.57227 |
| 13 | 1.76303 | 0.38151 |
| 14 | 1.50868 | 0.25435 |
| 15 | 1.33912 | 0.16956 |
| 16 | 1.22608 | 0.11304 |
| 17 | 1.15072 | 0.07536 |
| 18 | 1.10048 | 0.05024 |
| 19 | 1.06699 | 0.03349 |
| 20 | 1.04466 | 0.02233 |
| 21 | 1.02977 | 0.01489 |
| 22 | 1.01985 | 0.00992 |
| 23 | 1.01323 | 0.00662 |
| 24 | 1.00882 | 0.00441 |
| 25 | 1.00588 | 0.00294 |
| 26 | 1.00392 | 0.00196 |
| 27 | 1.00261 | 0.00131 |
| 28 | 1.00174 | 0.00087 |
| 29 | 1.00116 | 0.00058 |
| 30 | 1.00077 | 0.00039 |
| 31 | 1.00052 | 0.00025 |
| 32 | 1.00034 | 0.00018 |
| 33 | 1.00023 | 0.00011 |
| 34 | 1.00015 | 0.00008 |

CHAPTER 10 ■ ALGORITHMS FOR NUMERICAL ANALYSIS

# Integration

Another problem that frequently requires the help of mathematical algorithms is the integration of functions. The integral of a function can be visualized as the area under its graph, and it has many applications in finance, engineering, and physics. Several algorithms used in the analysis of options need to evaluate integrals numerically, using techniques similar to the ones covered in this section.

Functions can be integrated analytically or numerically. For some functions, it is possible to find an analytic solution, that is, a closed formula that can be directly evaluated to compute the integral of a function between two points. For example, polynomial functions can be easily integrated analytically, using the anti-derivative. For example, if the function is $f(x) = x^2$, the anti-derivative

$$F(x) = \frac{x^3}{3} + C$$

can be used to calculate the value of the integral between the two points a and b, which becomes $F(b) - F(a)$.

Many functions, however, are too complicated to be integrated analytically. In these cases, you need to use numerical algorithms that slice the function into small parts and calculate the integral, while trying to reduce the error in this process.

In this section, I present an implementation for one of the simplest integration techniques, known as Simpson's method. Simpson's method is based on the decomposition of an area that needs to be integrated into a large number of very small pieces.

First, you need to define a class that presents the interface for this solution method. The SimpsonsIntegration class contains data members such as m_f, a reference to the function that will be integrated, and m_numIntervals, the number of intervals used to approximate the integral.

```
#include "MathFunction.hpp"

class SimpsonsIntegration {
public:
    SimpsonsIntegration(MathFunction &f);
    SimpsonsIntegration(const SimpsonsIntegration &p);
    ~SimpsonsIntegration();
    SimpsonsIntegration &operator=(const SimpsonsIntegration &p);

    double getIntegral(double a, double b);
    void setNumIntervals(int n);
private:
    MathFunction &m_f;
    int m_numIntervals;
};
```

The implementation for this class is in the next code fragment. The class uses a default number of intervals, in case you don't want to set up this value. The DEFAULT_NUM_INTERVALS constant is used for this purpose.

```
#include "Integration.hpp"

#include "MathFunction.hpp"

#include <iostream>
#include <cmath>
```

CHAPTER 10 ■ ALGORITHMS FOR NUMERICAL ANALYSIS

```
using std::cout;
using std::endl;

namespace {
    const int DEFAULT_NUM_INTERVALS = 100;
}

SimpsonsIntegration::SimpsonsIntegration(MathFunction &f)
: m_f(f),
m_numIntervals(DEFAULT_NUM_INTERVALS)
{
}

SimpsonsIntegration::SimpsonsIntegration(const SimpsonsIntegration &p)
: m_f(p.m_f),
m_numIntervals(p.m_numIntervals)
{
}

SimpsonsIntegration::~SimpsonsIntegration()
{
}

SimpsonsIntegration &SimpsonsIntegration::operator=(const SimpsonsIntegration &p)
{
    if (this != &p)
    {
        m_f = p.m_f;
        m_numIntervals = p.m_numIntervals;
    }
    return *this;
}
```

The main part of this implementation is the getIntegral member function. The two parameters for this function define the interval in which the integration will be performed. The intSize variable is used to define the size of each interval used for Simpson's method.

The algorithm operates as follows. For each slice of the required interval, you need to compute the approximate area under the function. The formula used by Simpson's method is

$$\frac{b-a}{6}\left\{f(a)+4f\left(\frac{a+b}{2}\right)+f(b)\right\}$$

where a and b are the beginning and end points of the current interval. This rule has been observed as one of the most effective for evaluating an integral in a short interval.

```
double SimpsonsIntegration::getIntegral(double a, double b)
{
    double S = 0;
    double intSize = (b - a)/m_numIntervals;
    double x = a;
```

```
    for (int i=0; i<m_numIntervals; ++i)
    {
        S += (intSize / 6) * ( m_f(x) + m_f(x+intSize) + 4* m_f ((x + x+intSize)/2) );
        x += intSize;
    }
    return S;
}
```

This class also provides a method to change the number of intervals, therefore improving the accuracy of the method (at the expense of additional running time).

```
void SimpsonsIntegration::setNumIntervals(int n)
{
    m_numIntervals = n;
}
```

To test the results of this integration method, I provide a simple mathematical function as an example. The function to be integrated here is sin (x).

```
// Example function

namespace {

    class SampleFunc : public MathFunction
    {
    public:
        ~SampleFunc();
        double operator()(double x);
    };

    SampleFunc::~SampleFunc()
    {
    }

    double SampleFunc::operator()(double x)
    {
        return sin(x);
    }

}
```

The main function can be used as a driver to test the SimpsonsIntegration class. It creates an instance of SimpleFunc and uses it to initialize a SimpsonsIntegration object. Then, this code will call the function getIntegral for the interval 0.5 to 2.5. Next, the number of intervals changes to 200, and the same calculation is performed again.

```
int main()
{
    SampleFunc f;
    SimpsonsIntegration si(f);
    si.setNumIntervals(200);
```

```
        double integral = si.getIntegral(0.5, 2.5);
        cout << " the integral of the function is " << integral << endl;

        si.setNumIntervals(200);
        integral = si.getIntegral(0.5, 2.5);
        cout << " the integral of the function with 200 intervals is " << integral << endl;
        return 0;
}
```

The result of this function is the following:

**the integral of the function is 1.67876**
**the integral of the function with 200 intervals is 1.67873**

This is a very effective method, and with only four intervals, it possible to achieve a reasonable approximation in this case.

## Conclusion

Numerical algorithms are one of the main parts of an analytical system for options and derivatives. These algorithms have been refined for decades, and many of them have been implemented in C++ for the purpose of options pricing and related tasks.

In this chapter, you saw a few examples of numerical algorithms and learned how they can be efficiently implemented. I started with an explanation of how mathematical functions can be modeled as classes that are independent of the underlying algorithm. You also saw how to create a generic polynomial function class that efficiently computes the value of a function at each point using Horner's method.

Next, you learned how to find roots of equations using Newton's method. This traditional method employs the derivative of a function to estimate the value of the root, and continually improves this estimate until a solution is found. You saw how this method can be relatively easily implemented using the tools developed in the previous sections.

Finally, this chapter also covered the important problem of function integration. To find the integral of a function, you need to evaluate a function in a given range and use those values to estimate the area covered by the function graph. Using the algorithmic methods introduced here, you saw how to implement one of the most common techniques for integrating functions, known as Simpson's method.

While this chapter introduced simple numeric techniques, in the next chapter you will learn how these techniques can be combined to solve some of the complex differential equations that are common when analyzing options and related derivatives.

# CHAPTER 11

# Models Based on Differential Equations

Differential equations are equations that involve in their terms both a function as well as their mathematical derivatives. Many of these equations arise naturally from the analysis of economic models used for the pricing of options, such as the Black-Scholes model.

Solving specific partial differential equations (PDEs) is at the core of many techniques used in the analysis of options and related financial derivatives. As you will see in this chapter, there are several techniques for solving and analyzing the results of PDEs that can be implemented in C++. In the next few sections, I present programming examples that cover important aspects of differential equations-based option modeling and their applications using C++.

Here are a few of the topics covered in this chapter:

- *Basic techniques for solving DEs*: Several techniques have been developed by practitioners in order to find solutions for differential equations. I provide a quick summary of these methods and explain how they can be used in financial applications.

- *Ordinary differential equations*: ODEs are equations that contain only functions and derivatives of one value. ODEs can be used to represent problems in several areas, and solving them gives you an excellent basis for solving more complex differential equations.

- *Euler's method for solving ODEs*: Euler's method is a traditional algorithm that can be easily implemented in C++, providing a numerical evaluation method for a large number of DEs.

- *Runge-Kutta method*: The RK method provides a more accurate way to determine numerical solutions for differential equations. The RK method uses a Taylor expansion as a way to approximate the desired equation, which makes it possible to find solutions with fewer iterations of the algorithm.

## General Differential Equations

Differential equations (DEs) are defined as equations that include one or more derivatives of a function. They have an important role in modeling several types of phenomena occurring in diverse areas such as physics, engineering, social sciences, and economy. In physics, for instance, differential equations are typically used to model the dynamics of motion and forces. In economics, it is possible to use DEs to model financial systems that involve interest rates and time decay.

CHAPTER 11 ■ MODELS BASED ON DIFFERENTIAL EQUATIONS

Differential equations are very useful because they encode information about the rate of variation of a particular quantity. The derivative is the concept that represents the rate of change of a function with respect to a particular variable. The second derivative, in its turn, represents the rate of change of the first derivative with respect to the original variable. The same strategy can be used for as many derivatives as needed by the application.

Differential equations are classified according to the terms they contain, involving functions and their derivatives. Here are some examples of differential equations:

$$\frac{dy}{dx} + x^2 y = 2x$$

This is a differential equation involving quantities $x$ and $y$, with a first derivative of $y$ with respect to $x$ and a few other standard terms.

$$10x\frac{d^2y}{dx^2} + x\frac{dy}{dx} = 0$$

This is a differential equation that involves a second derivative of $y$ with respect to $x$, as well as the first derivative.

The *order* of a differential equation is the maximum order of the derivatives appearing in it. For example, a first-order differential equation includes only the first derivative. A second-order differential equation may also contain second-order derivatives, such as

$$\frac{d^2x}{dt^2}$$

To solve differential equations, it is frequently useful to separate them into particular categories and develop solution techniques that can handle such specific categories. In the next sections, you will see specific types of DEs as well as some solution techniques developed for these types of equations.

## Ordinary Differential Equations

An ordinary differential equation is a type of DE in which functions of only a single (ordinary) variable are allowed to appear. As with other types of differential equations, ODEs include variables, functions, and their derivatives. A formal definition of an ODE is a function

$$F(x, f(x), f'(x), f''(x), \ldots, f^n(x))$$

that depends on a variable $x$, a function $f(x)$ of $x$, and their derivatives. The order of the ODE is the maximum order of derivatives appearing in the equation.

You can solve ODEs in two ways:

- *Using analytical methods*: If the function can be solved explicitly using mathematical methods, then a closed expression can be found and used to calculate its value at different points. This method is preferred whenever possible, because it produces results that are usually easier to calculate and interpret. Unfortunately, it is not always possible to find closed solutions to complex differential equations.

- *Using numerical methods*: More generally, it is difficult to find closed solutions for several classes of ODEs. In this case, the analyst may resort to using numerical techniques that approximate the value of the ODE for a particular value or range of values. These numerical techniques usually involve the approximation of the value of a complex function in a piecewise fashion, so that the solution of the differential equation is found after a large number of small approximation steps.

CHAPTER 11 ■ MODELS BASED ON DIFFERENTIAL EQUATIONS

Since the goal of this chapter is to consider computational techniques to solve ODEs, you will see a few techniques to solve them numerically, using programming strategies. First, you will learn about Euler's method for solving ODEs. Then, you will see how this method can be implemented in C++.

## Euler's Method

One of the most common methods used for solving ODEs is called Euler's method. It was one of the first algorithms developed for this purpose, and was proposed by the famous XVIII century mathematician Leonard Euler. The method belongs to a class of ODE algorithms called predictor-corrector, because it tries to make a prediction for the next step in the evaluation, followed by successive corrections of the current result.

The basic idea behind Euler's is to approximate a curve determined by a differential equation through sequential steps. First, to start the solution process, you need to represent the ODE in its most generic form:

$$y' = F(x,y)$$

Here, $y = f(x)$ is a function that depends on the variable $x$, and $y'$ is the derivative of $f(x)$ with respect to $x$. The general goal of the method is to improve the approximation step by step, using a simple formula to calculate small increments and using the result as the next starting point. Figure 11-1 shows an example of how the general approach works, when applied to the sample differential equation $dT(t)/dt = -k\,\Delta T$.

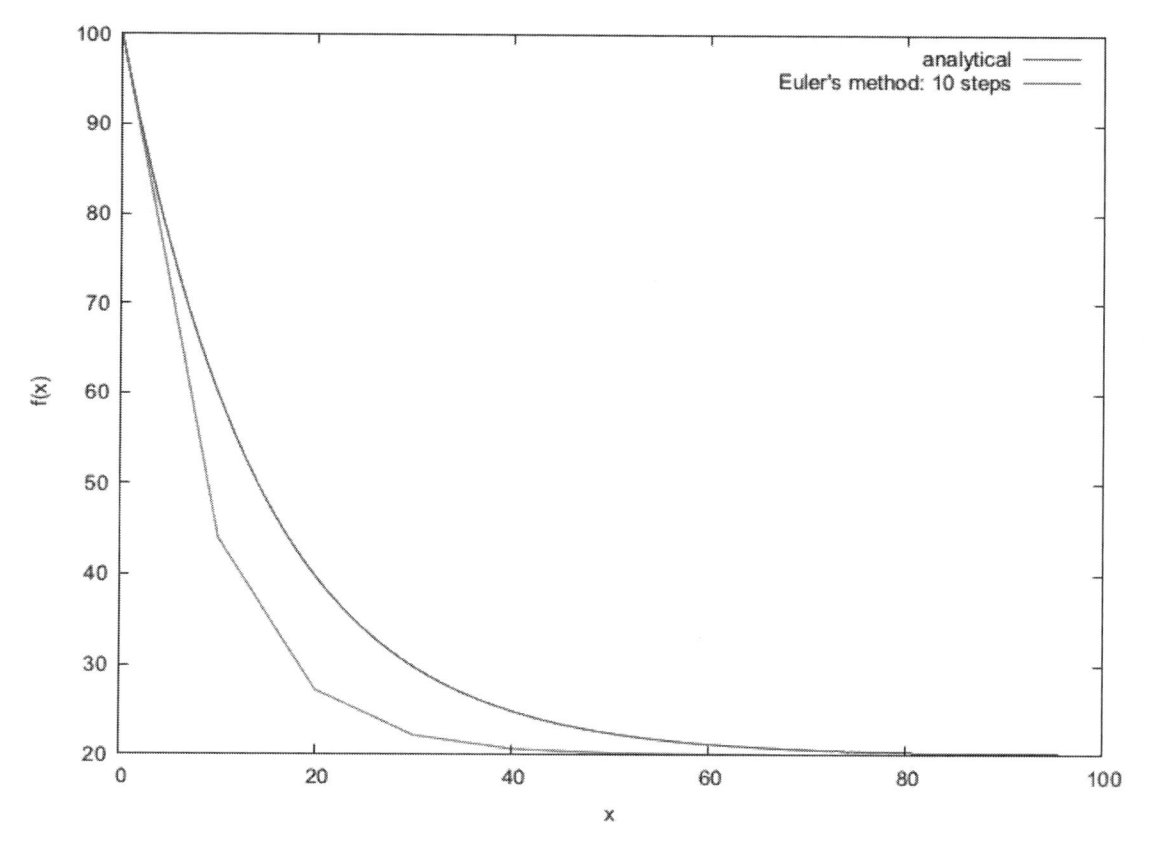

**Figure 11-1.** *Euler's method applied to function $dT(t)/dt = -k\,\Delta T$, with 10 steps*

Each step starts at a known place of the solution space and moves into the required direction by a small quantity. If you denote by $c$ the desired destination point and start moving from location $x_0$ in $N$ steps, then the increment h can be calculated as follows

$$h = \frac{(c - x_0)}{N}$$

Now, at each step of this algorithm, you will have the current location (at the beginning the location is $(x_0, y_0)$, a given parameter passed to the algorithm) and the goal is to compute the next location that approximates the real curve. As long as h is small enough, this new location can be calculated by taking the derivative of the curve, given by $y'$, which represents the slope of the equation, and using a simple line segment to move in that direction. This is fairly easy to calculate numerically, as you will see next.

The equation needed to implement this idea is the following:

$$y_t = y_{t-1} + h \frac{f(x_{t-1}, y_{t-1}) + f(x_{t-1} + h, y + hf(x_{t-1}, y_{t-1}))}{2}$$

In other words, at each step, you're adding to the previous result a quantity that depends on the step size and the average value of the target function at two points: the current point and the next incremental point. You can think of the averaging (dividing by two) as a correction of the procedure, which will make it closer to the real value that needs to be computed.

## Implementing the Method

Euler's method can be implemented with little effort. First, you need to update the MathFunction class so that it can also be used when a variable and an initial condition are provided. This requires that the function call operator take two parameters instead of one, such as was presented in the last chapter. I coded this as a class called DEMathFunction, with this interface:

```
class DEMathFunction : MathFunction {
public:

    virtual ~DEMathFunction() {}
    virtual double operator()(double x, double y) = 0; // version with two variables
private:
    // this just an interface
};
```

The new version of operator() takes as parameters the value of coordinates $x$ and $y$. Now, you can implement versions of this class for each desired function. Here is an example that will later be used with the main implementation:

```
class EulerMethodSampleFunction : public DEMathFunction {
public:
    double operator()(double x, double y);
};

double EulerMethodSampleFunction::operator()(double x, double y)
{
    return  3 * x + 2 * y + 1;
}
```

## CHAPTER 11 ■ MODELS BASED ON DIFFERENTIAL EQUATIONS

The main class implementing Euler's method is presented next. The interface contains a single function called solve, which receives four parameters:

- The number of steps used by the algorithm
- The initial x value
- The initial y value (which represents the initial condition of the function)
- The target value for the ODE, which is the coordinate for which the solution is required

The class also contains a data member to store the instance of DEMathFunction, which is used to compute new values for the desired function.

```
class EulersMethod {
public:
    EulersMethod(DEMathFunction &f);
    EulersMethod(const EulersMethod &p);
    ~EulersMethod();
    EulersMethod &operator=(const EulersMethod &p);

    double solve(int n, double x0, double y0, double c);
private:
    DEMathFunction &m_f;
};
```

The implementation of the EulersMethod class contains the steps of the algorithm explained in previous section. First, here are some of the required methods used by the class:

```
//
//   EulersMethod.cpp

#include "EulersMethod.hpp"

#include <iostream>

using std::cout;
using std::endl;

EulersMethod::EulersMethod(DEMathFunction &f)
: m_f(f)
{
}

EulersMethod::EulersMethod(const EulersMethod &p)
: m_f(p.m_f)
{
}

EulersMethod::~EulersMethod()
{
}
```

179

## CHAPTER 11 ■ MODELS BASED ON DIFFERENTIAL EQUATIONS

```
EulersMethod &EulersMethod::operator=(const EulersMethod &p)
{
    if (this != &p)
    {
        m_f = p.m_f;
    }
    return *this;
}
```

Next, the solve function contains the main algorithm for Euler's method. The algorithm assumes that x0 is the initial coordinate and y0 is the corresponding initial value for that coordinate.

```
double EulersMethod::solve(int n, double x0, double y0, double c)
{
    // problem :   y' = f(x,y) ;  y(x0) = y0

    auto x = x0;
    auto y = y0;
    auto h = (c - x0)/n;

    cout << " h is " << h << endl;

    for (int i=0; i<n; ++i)
    {
        double F = m_f(x, y);
        auto G = m_f(x + h, y + h*F);

        cout << " F: " << F << " G: " << G << "";

        // update values of x, y
        x += h;
        y += h * (F + G)/2;

        cout << " x: " << x << " y: " << y << endl;
    }

    return y;
}
```

The first part of the algorithm uses the given values to calculate the desired increment h. Then, for each step the algorithm will calculate the function at the current point (x,y), as well as at the next incremental point (x + h, y + hF). The values of x and y are then updated according to the equation presented in the previous section.

You can quickly test the implementation with the help of the EulerMethodSampleFunction class. Here is the sample code necessary to instantiate the class and use it to test the method:

```
int test_euler()
{
    EulerMethodSampleFunction f;
    EulersMethod m(f);
    double res = m.solve (100, 0, 0.25, 2);
    cout << " result is " << res << endl;
    return 0;
}
```

The sample function is instantiated in the first line, and the resulting function object is passed to EulersMethod class. The member function solve is called, with a few initial parameters. The results are printed as the last step. Table 11-1 shows the sequence of values obtained when you run the test function.

*Table 11-1. Results of Euler's Method Iterations for the Test Code for the EulersMethod Class*

| i | F | x | y | i | F | x | y | i | F | x | y |
|---|---|---|---|---|---|---|---|---|---|---|---|
| 1 | 1.5 | 0.02 | 0.2812 | 34 | 9.72643 | 0.68 | 3.57223 | 67 | 40.5109 | 1.34 | 18.6025 |
| 2 | 1.6224 | 0.04 | 0.314897 | 35 | 10.1845 | 0.7 | 3.7806 | 68 | 42.2249 | 1.36 | 19.4645 |
| 3 | 1.74979 | 0.06 | 0.351193 | 36 | 10.6612 | 0.72 | 3.99868 | 69 | 44.0089 | 1.38 | 20.3628 |
| 4 | 1.88239 | 0.08 | 0.390193 | 37 | 11.1574 | 0.74 | 4.2269 | 70 | 45.8657 | 1.4 | 21.2991 |
| 5 | 2.02039 | 0.1 | 0.432009 | 38 | 11.6738 | 0.76 | 4.46564 | 71 | 47.7982 | 1.42 | 22.2748 |
| 6 | 2.16402 | 0.12 | 0.476755 | 39 | 12.2113 | 0.78 | 4.71535 | 72 | 49.8096 | 1.44 | 23.2915 |
| 7 | 2.31351 | 0.14 | 0.524551 | 40 | 12.7707 | 0.8 | 4.97647 | 73 | 51.903 | 1.46 | 24.3509 |
| 8 | 2.4691 | 0.16 | 0.575521 | 41 | 13.3529 | 0.82 | 5.24947 | 74 | 54.0818 | 1.48 | 25.4548 |
| 9 | 2.63104 | 0.18 | 0.629794 | 42 | 13.9589 | 0.84 | 5.53484 | 75 | 56.3496 | 1.5 | 26.6049 |
| 10 | 2.79959 | 0.2 | 0.687505 | 43 | 14.5897 | 0.86 | 5.83306 | 76 | 58.7098 | 1.52 | 27.8032 |
| 11 | 2.97501 | 0.22 | 0.748796 | 44 | 15.2461 | 0.88 | 6.14469 | 77 | 61.1664 | 1.54 | 29.0516 |
| 12 | 3.15759 | 0.24 | 0.81381 | 45 | 15.9294 | 0.9 | 6.47025 | 78 | 63.7232 | 1.56 | 30.3522 |
| 13 | 3.34762 | 0.26 | 0.882702 | 46 | 16.6405 | 0.92 | 6.81031 | 79 | 66.3843 | 1.58 | 31.707 |
| 14 | 3.5454 | 0.28 | 0.955628 | 47 | 17.3806 | 0.94 | 7.16548 | 80 | 69.154 | 1.6 | 33.1183 |
| 15 | 3.75126 | 0.3 | 1.03275 | 48 | 18.151 | 0.96 | 7.53636 | 81 | 72.0367 | 1.62 | 34.5885 |
| 16 | 3.96551 | 0.32 | 1.11425 | 49 | 18.9527 | 0.98 | 7.92359 | 82 | 75.037 | 1.64 | 36.1198 |
| 17 | 4.1885 | 0.34 | 1.2003 | 50 | 19.7872 | 1 | 8.32785 | 83 | 78.1597 | 1.66 | 37.7149 |
| 18 | 4.42059 | 0.36 | 1.29108 | 51 | 20.6557 | 1.02 | 8.74983 | 84 | 81.4098 | 1.68 | 39.3763 |
| 19 | 4.66215 | 0.38 | 1.38678 | 52 | 21.5597 | 1.04 | 9.19024 | 85 | 84.7925 | 1.7 | 41.1066 |
| 20 | 4.91357 | 0.4 | 1.48762 | 53 | 22.5005 | 1.06 | 9.64985 | 86 | 88.3132 | 1.72 | 42.9088 |
| 21 | 5.17524 | 0.42 | 1.5938 | 54 | 23.4797 | 1.08 | 10.1294 | 87 | 91.9776 | 1.74 | 44.7858 |
| 22 | 5.44759 | 0.44 | 1.70553 | 55 | 24.4989 | 1.1 | 10.6298 | 88 | 95.7915 | 1.76 | 46.7405 |
| 23 | 5.73105 | 0.46 | 1.82304 | 56 | 25.5596 | 1.12 | 11.1518 | 89 | 99.761 | 1.78 | 48.7762 |
| 24 | 6.02608 | 0.48 | 1.94657 | 57 | 26.6637 | 1.14 | 11.6964 | 90 | 103.892 | 1.8 | 50.8962 |
| 25 | 6.33314 | 0.5 | 2.07637 | 58 | 27.8127 | 1.16 | 12.2643 | 91 | 108.192 | 1.82 | 53.104 |
| 26 | 6.65274 | 0.52 | 2.21268 | 59 | 29.0087 | 1.18 | 12.8567 | 92 | 112.668 | 1.84 | 55.403 |
| 27 | 6.98537 | 0.54 | 2.35578 | 60 | 30.2535 | 1.2 | 13.4745 | 93 | 117.326 | 1.86 | 57.797 |
| 28 | 7.33157 | 0.56 | 2.50595 | 61 | 31.549 | 1.22 | 14.1187 | 94 | 122.174 | 1.88 | 60.29 |
| 29 | 7.6919 | 0.58 | 2.66346 | 62 | 32.8974 | 1.24 | 14.7904 | 95 | 127.22 | 1.9 | 62.8859 |
| 30 | 8.06693 | 0.6 | 2.82863 | 63 | 34.3008 | 1.26 | 15.4907 | 96 | 132.472 | 1.92 | 65.5889 |
| 31 | 8.45726 | 0.62 | 3.00176 | 64 | 35.7615 | 1.28 | 16.2209 | 97 | 137.938 | 1.94 | 68.4034 |
| 32 | 8.86351 | 0.64 | 3.18317 | 65 | 37.2817 | 1.3 | 16.982 | 98 | 143.627 | 1.96 | 71.334 |
| 33 | 9.28635 | 0.66 | 3.37321 | 66 | 38.864 | 1.32 | 17.7754 | 99 | 149.548 | 1.98 | 74.3854 |

CHAPTER 11 ■ MODELS BASED ON DIFFERENTIAL EQUATIONS

Euler's method is a simple technique that finds solutions to several ODE problems. However, in terms of quality of approximation, it requires a large number of steps, which can also cause numerical errors and instability. To avoid these problems, more precise methods have been proposed for solving ODEs, as you will learn next.

# The Runge-Kutta Method

The next technique for solving ODEs is an extension of Euler's method called the Runge-Kutta (RK) method (named after its inventors). This technique is an effective way to improve the accuracy of Euler's method and reduce the possibility of the numerical errors that are common when using a linear approximation.

The main idea of the RK method is to use a higher-order approximation for the given functions, instead of relying on linear interpolation, as you saw with the previous algorithm. By doing this, the RK method can achieve faster convergence, in many cases using a smaller number of steps to achieve the same results. This is an advantage both in terms of reduced computational time as well as higher accuracy.

Remember that to solve an ODE, you have to consider a very general form that is amenable to solution, using the following relation:

$$y' = f(x,y)$$

Here, $y'$ is the derivative of the function and $f(x,y)$ is a function of variable $x$ (the independent variable) and $y$.

As before, given a starting point for the calculation and the number of steps, it is possible to easily calculate the size of the increment $h$ for each iteration of the RK method, using the equation

$$h = \frac{c - x_0}{N}$$

In its basic design, the RK method has the same structure of Euler's algorithm. The main difference is how the RK method approximates the function to generate the next step of the algorithm. While Euler's method just uses a linear interpolation, the RK method can use any one of a family of approximating equations.

The RK method can be implemented using one of several approximation strategies, but they are frequently calculated as a Taylor series applied to the original function. The Taylor method is a basic tool from calculus that provides a family of approximations for functions around a particular starting value. For example, using the simplest Taylor approximation, you can compute the next $(x,y)$ values in the following way:

$$x_{t+1} = x_t + h$$

$$y_{t+1} = y_t + hf\left(x_t + \frac{h}{2}, y_t + \frac{h}{2}f(x_t, y_t)\right)$$

Another possibility is to use higher-order approximations, that is, versions of the Taylor series that contain additional terms. By adding more terms of higher order, it is possible to achieve a more accurate result in fewer steps. Here is another commonly used approximation, this time based on a fourth-order expansion:

$$k_1 = hf(x_t, y_t)$$

$$k_2 = hf\left(x_t + \frac{h}{2}, y_t + \frac{k_1}{2}\right)$$

$$k_3 = hf\left(x_t + \frac{h}{2}, y_t + \frac{k_2}{2}\right)$$

$$k_4 = hf(x_t + h, y_t + k_3)$$

$$y_{t+1} = y_t + \frac{1}{6}(k_1 + 2k_2 + 2k_3 + k_4)$$

# Runge-Kutta Implementation

To implement this algorithm, it is possible to extend the Euler's method class. To avoid dependencies between these two methods, I decided to implement a separate class called RungeKuttaMethod.

Here is the interface of the RungeKuttaMethod class. It exposes the solve method, which is used to compute the desired value of the function.

```
//
//   class providing an interface for RungeKutta method

class RungeKuttaMethod {
public:
    RungeKuttaMethod(DEMathFunction &f);
    RungeKuttaMethod(const RungeKuttaMethod &p);
    ~RungeKuttaMethod();
    RungeKuttaMethod &operator=(const RungeKuttaMethod &p);
    double solve(int n, double x0, double y0, double c);
private:
    DEMathFunction &m_func;
};
```

First, the common member functions of RungeKuttaMethod are implemented, including the constructor that receives the DEMathFunction reference as a parameter.

```
//
//   RungeKutta.cpp

#include "RungeKutta.hpp"

#include <iostream>

using std::cout;
using std::endl;

RungeKuttaMethod::RungeKuttaMethod(DEMathFunction &f)
: m_func(f)
{
}

RungeKuttaMethod::RungeKuttaMethod(const RungeKuttaMethod &p)
: m_func(p.m_func)
{
}

RungeKuttaMethod::~RungeKuttaMethod()
{
}
```

## CHAPTER 11 ▪ MODELS BASED ON DIFFERENTIAL EQUATIONS

```
RungeKuttaMethod &RungeKuttaMethod::operator=(const RungeKuttaMethod &p)
{
    if (this != &p)
    {
        m_func = p.m_func;
    }
    return *this;
}
```

The member function solve is used to compute the numerical value of the ODE, given starting conditions and a target value. The function implements the Runge-Kutta method with fourth-degree Taylor expansion, as described in the previous section.

The parameters for this member function are the following:

- The number of steps in the process, which indirectly also determines the increment for each step
- The initial value for the variable $x$.
- The initial corresponding $y$ for the given value $x$.
- The target value for which the ODE is being calculated

```
// Runge-Kutta method with fourth order approximation
//
double RungeKuttaMethod::solve(int n, double x0, double y0, double c)
{
    // initial conditions
    auto x = x0;
    auto y = y0;
    auto h = (c - x0)/n;

    for (int i=0; i<n; ++i)
    {
        // compute the intermediary steps
        //
        auto k1 = h * m_func(x, y);
        auto k2 = h * m_func(x + (h/2), y + (k1/2));
        auto k3 = h * m_func(x + (h/2), y + (k2/2));
        auto k4 = h * m_func(x + h, y + k3);

        // use terms to compute next step
        x += h;
        y += ( k1 + 2*k2 + 2*k3 + k4)/6;
        cout << " x: " << x << " y: " << y << endl;
    }

    return y;
}
```

As in the previous algorithm, the RK method starts by defining the initial conditions, including the values for the variables $x$ and $y$, and the size of the step determined by h.

The RK method then proceeds to compute each iteration of the algorithm. This consists of successive terms of approximation, as described in the previous section. These terms are then used to compute the new values for $x$ and $y$.

To test the results of the RK method implementation, I provide a simple test function. But first it is necessary to implement a function that will be later used in the test code:

```
class RungeKuttaSampleFunc : public DEMathFunction {
public:

    double operator()(double x, double y);
};

double RungeKuttaSampleFunc::operator()(double x, double y)
{
    return  3 * x + 2 * y + 1;
}
```

The RungeKuttaSampleFunc is derived from DEMathFunction, so it can be passed as a parameter to the RungeKuttaMethod class. It is a simple polynomial function. The test function is the following:

```
int test_RKMethod()
{
    RungeKuttaSampleFunc f;
    RungeKuttaMethod m(f);
    double res = m.solve (100, 0, 0.25, 2);
    cout << " result is " << res << endl;
    return 0;
}
```

This test code first instantiates the RungeKuttaSampleFunc class and then uses the resulting instance to create a RungeKuttaMethod object. Next, the result of the function is computed for some test parameters.

## Complete Code

The complete listing for the RungeKuttaMethod class is shown in this section. The code is divided into a header file and an implementation file, which appear in Listings 11-1 and 11-2, respectively.

***Listing 11-1.*** Header File for the RungeKuttaMethod Class

```
//
//  RungeKutta.hpp

#ifndef RungeKutta_hpp
#define RungeKutta_hpp

#include "EulersMethod.hpp"

class RungeKuttaMethod {
public:
    RungeKuttaMethod(DEMathFunction &f);
    RungeKuttaMethod(const RungeKuttaMethod &p);
    ~RungeKuttaMethod();
    RungeKuttaMethod &operator=(const RungeKuttaMethod &p);
    double solve(int n, double x0, double y0, double c);
```

CHAPTER 11 ■ MODELS BASED ON DIFFERENTIAL EQUATIONS

```
private:
    DEMathFunction &m_func;
};

#endif /* RungeKutta_hpp */
```

***Listing 11-2.*** Implementation File for the RungeKuttaMethod Class

```cpp
//
//  RungeKutta.cpp

#include "RungeKutta.hpp"

#include <iostream>

using std::cout;
using std::endl;

RungeKuttaMethod::RungeKuttaMethod(DEMathFunction &f)
: m_func(f)
{
}

RungeKuttaMethod::RungeKuttaMethod(const RungeKuttaMethod &p)
: m_func(p.m_func)
{
}

RungeKuttaMethod::~RungeKuttaMethod()
{
}

RungeKuttaMethod &RungeKuttaMethod::operator=(const RungeKuttaMethod &p)
{
    if (this != &p)
    {
        m_func = p.m_func;
    }
    return *this;
}

// Runge-Kutta method with fourth order approximation
//
double RungeKuttaMethod::solve(int n, double x0, double y0, double c)
{
    // initial conditions
    auto x = x0;
    auto y = y0;
    auto h = (c - x0)/n;
```

```cpp
    for (int i=0; i<n; ++i)
    {
        // compute the intermediary steps
        //
        auto k1 = h * m_func(x, y);
        auto k2 = h * m_func(x + (h/2), y + (k1/2));
        auto k3 = h * m_func(x + (h/2), y + (k2/2));
        auto k4 = h * m_func(x + h, y + k3);

        // use terms to compute next step
        x += h;
        y += ( k1 + 2*k2 + 2*k3 + k4)/6;
        cout << " x: " << x << " y: " << y << endl;
    }

    return y;
}

/// -----

class RungeKuttaSampleFunc : public DEMathFunction {
public:

    double operator()(double x, double y);
};

double RungeKuttaSampleFunc::operator()(double x, double y)
{
    return  3 * x + 2 * y + 1;
}

int main_rkm()
{
    RungeKuttaSampleFunc f;
    RungeKuttaMethod m(f);
    double res = m.solve (100, 0, 0.25, 2);
    cout << " result is " << res << endl;
    return 0;
}
```

## Conclusion

Solving differential equations is a task commonly required when analyzing complex financial contracts. This is true due to the mathematical nature of options and derivatives, which are based on the Black-Scholes model.

In this chapter, you saw a few examples of differential equations, and learned how the can be effectively solved using computational techniques. First, you learned about Euler's method, the simplest technique used to compute numerical solutions for ODEs. Next, you learned about the Runge-Kutta method, a commonly used technique that provides improved accuracy over Euler's method.

This chapter can be used as an overview of the implementation of differential equations in C++. In the next chapter, you will take a closer look at how these mathematical models can be directly applied to option pricing. In particular, you will see how these techniques can be used when pricing option contracts.

# CHAPTER 12

# Basic Models for Options Pricing

Options pricing is the task of determining the fair value of a particular option, given a set of parameters that exactly determine the features of the option contract, such as its expiration date, current volatility, and prevailing interest rates. Pricing options requires the use of efficient algorithms, because of frequent changes in prices and market volatility. For this reason, a number of models have been employed for this task in the area of quantitative finance.

This chapter discusses some of the most popular models for options pricing. First, there are models that use tree-based methods, such as binomial and trinomial trees. Second, the most important mathematical model uses the Black-Scholes model, which provides the theoretical basis for the analysis of most options and derivative contracts.

Here is a summary of the topics discussed in this chapter:

- *Binomial trees*: A binomial tree is a technique used to compute option prices by simulating a number of probabilistic price changes starting from the current stock price. Such prices are organized in a tree-based structure and used to compute the option's corresponding price. You will see the calculations necessary to use these tree-based algorithms for options pricing.

- *Calculating American-style options*: Options in the American style give their buyers the ability to exercise the option at any time before expiration. This exercise style needs to be reflected in the price of the option.

- *Black-Scholes method*: The most famous method for computing option prices is based on the equations developed by Black and Scholes. These differential equations can be solved using PDE techniques, which are explored later in this chapter.

- *Implementation strategies*: You will see examples of implementation techniques for the pricing methods described previously.

## Lattice Models

The goal of options pricing is to compute the fair value of an option at a particular time. This problem has been solved theoretically by Black and Scholes, the creators of the famous PDE model that defines prices for options. However, solving complex PDEs is not an easy job, and for this reason several methods have been developed to perform this computational task in less time.

A common class of algorithms for computing options prices is the *lattice model*. A lattice model is a technique of calculating derivative prices that divides the solution space into discrete steps. Each step corresponds to a small time increment and corresponding price change. Starting this way from a given starting point, this technique results in the creation of a tree of nodes that corresponding to possible price changes.

CHAPTER 12 ■ BASIC MODELS FOR OPTIONS PRICING

There are a few particular methods that have been devised based on the general strategy put forward by lattice models. The best known such methods are:

- *Binomial model*: In the binomial model, the possible changes are organized in a tree rooted at the given starting point (the current price). To each node of the tree, two nodes are added representing two possible directions of movement: up (price increases) or down (price decreases). For performance reasons, the binary tree can also be created implicitly, where nodes are calculated only as needed for the evaluation of the next time period.

- *Trinomial model*: The trinomial model is an extension of the binomial model and it tries to improve the accuracy by considering nodes where the price is unchanged. Depending on the volatility of the underlying, such models can achieve higher accuracy than binomial models, at the expense of a slight increase in computational time.

Mixed models have also been used that combine features of the binomial and trinomial models, producing more complex lattice models for particular uses. In this chapter, you learn how to implement a binomial model for options pricing. The complete model is explained along with the equations frequently used to evaluate such models.

Later, this general model is extended to handle American-style options, where the owners of the option can exercise the option at any time before expiration. These models also show how this type of algorithm can be efficiently coded in C++ using OO concepts. In this particular case, you will see how to use inheritance to override parts of the class according to the desired pricing strategy.

## Binomial Model

The first model that's discussed is called the binomial model for options pricing. In this model, options prices are evaluated interactively. Possible values are organized in a tree-based structure where the root is the original (unknown) price and leaves are the possible prices at a particular target time.

Using this structure, the binomial model traverses the tree with the goal of computing the desired price (the root value) starting from some known prices. The natural way of doing this is to look at the values for the option at expiration date and use these prices to compute the value at other times. Remember that at expiration price, the value of an option is defined by contract. For example, if you are given the current stock price (denoted by S) and the strike price (denoted by K), then the price of a call option at expiration is given by:

$$p_c(S) = \max(0, S-K)$$

For a put option, the price is also straightforward and determined by contract as:

$$p_p(S) = \max(0, K-S)$$

The question, however, is which values of stock prices should be used in a tree-based model to make it realistic? A possible answer to this question is that at each time step, the stock price can move either up or down. The exact probabilities for this jump can be derived using a few mathematical assumptions, but the expressions most commonly used are as follows.

- Change of value for an up move:

$$\exp(\sigma\sqrt{t})$$

- Change of value for a down move:

$$\exp(-\sigma\sqrt{t})$$

In these two expressions, σn is a measure of the volatility of the stock (i.e., the typical amount of movement) and t is time. These expressions allow you to construct a tree where each node contains information about the time and the value of the stock at that moment. The tree can be visualized as shown in Figure 12-1.

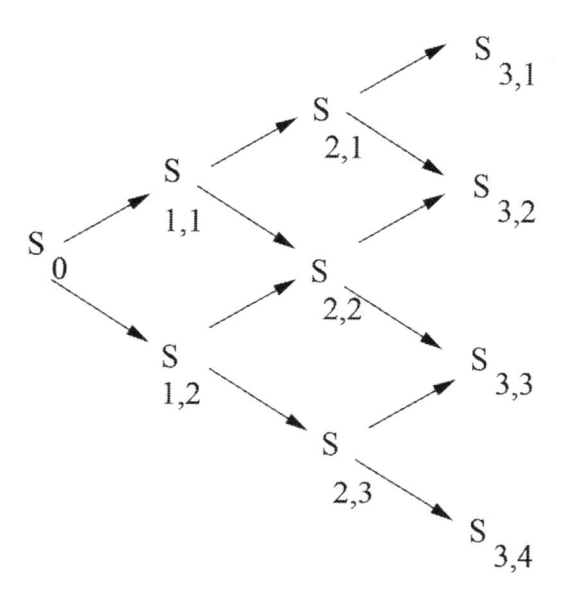

*Figure 12-1. A visualization of the binomial tree determined by possible stock prices*

Now consider the task of pricing a call option at a date immediately before expiration. While the price is initially unknown, it cannot be very far away from the price at expiration, since the time premium at this point is very small. A way to calculate this value is to assume a probability for two events: either going up a small amount or going down a small amount. With this probability, you can estimate the value of the option as the expected value (the mean) based on these two possibilities.

Using these observations, you can devise a method for calculating the price of an option. The general algorithm can be described in the following way.

- Calculate stock prices for the nodes of the tree, starting from the root node at time zero and stock price given by the current known price.

- Apply the equations for price fluctuations to create up and down nodes starting from the root. The goal of this phase is to calculate the stock prices for nodes at expiration time.

- Start to compute the option prices from the leaves of the tree. These leaves have a known price by definition of the option contract. The value of the option depends on three characteristics:

    - The strike price
    - The stock price
    - If the option is a put or a call

- Then, progress from nodes at expiration date toward earlier dates, always using the expected value based on the known probabilities. Repeat this process until you reach the root node.

CHAPTER 12 ■ BASIC MODELS FOR OPTIONS PRICING

# Binomial Model Implementation

To implement an algorithm for the binomial model as previously described, I introduce a class called BinomialModel. The class provides all the necessary steps for the calculation of option prices, along with the ability to be extended to other open types, as you will see later.

The first step is to provide an interface to the C++ class, as shown in the next code fragment. The class contains a number of data members that are necessary for the computation of options prices using the binomial model approach. Here are these data members:

- The expiration date, denoted as m_T.

- The initial stock price, that is, the stock price at the root of the binomial tree, denoted by m_S.

- The interest rate, which is used as one of the factors necessary to calculate future prices, and is denoted as m_r.

- The volatility, which is the volatility of the underlying stock, as measured from stock prices in the last few days and denoted by m_sigma.

- The dividend yield, which is the amount of dividend paid by the underlying stock during the desired period. This quantity is denoted by m_q.

- The number of steps, used by the binomial method to determine the depth of the tree. It is denoted by m_n.

- The type of option. This is the class record if the option type is a call or put. This information is stored in the member variable m_call, a Boolean value.

The class BinomialModel also offers a member function that can be used to calculate the option price, named optionPriceForStrike. This function receives as a parameter a strike value and returns the option price corresponding to that strike.

A second function, computePriceStep, is used to compute option prices for a single step. You will see later how this is implemented and extended for more complex option types.

```
class BinomialModel {
public:
    BinomialModel(const BinomialModel &p);
    virtual ~BinomialModel();
    BinomialModel &operator=(const BinomialModel &p);

    BinomialModel(double T, // expiration time
                double S,   // stock price
                double r,   // interest rate
                double sigma,
                double q,   // dividend yield
                int n,      // number of steps
                bool call
    );

    double optionPriceForStrike(double K);
    virtual void computePriceStep(int i, int j, double K, vec &prices,
                            double p_u, double p_d, double u);
```

## CHAPTER 12 ■ BASIC MODELS FOR OPTIONS PRICING

```
    protected:
        double getStockPrice() { return m_S; }
    private:
        double m_T;       // expiration time
        double m_S;       // stock price
        double m_r;       // interest rate
        double m_sigma;   // volatility
        double m_q;       // dividend yield
        int m_n;          // number of steps
        bool m_call;      // true = call, false = put
};
```

The next few member functions are part of the constructor and destructor code. They are used to properly initialize each of the data members in the `BinomialModel` class.

```
BinomialModel::BinomialModel(double T, double S, double r,
            double sigma,
            double q,
            int n, bool call)
: m_T(T),
  m_S(S),
  m_r(r),
  m_sigma(sigma),
  m_n(n),
  m_q(q),
  m_call(call)
{
}

BinomialModel::BinomialModel(const BinomialModel &p)
: m_T(p.m_T),
  m_S(p.m_S),
  m_r(p.m_r),
  m_sigma(p.m_sigma),
  m_n(p.m_n),
  m_q(p.m_q),
  m_call(p.m_call)
{
}

BinomialModel::~BinomialModel()
{
}

BinomialModel &BinomialModel::operator=(const BinomialModel &p)
{
    if (this != &p)
    {
```

# CHAPTER 12 ■ BASIC MODELS FOR OPTIONS PRICING

```
        m_T = p.m_T;
        m_S = p.m_S;
        m_r = p.m_r;
        m_sigma = p.m_sigma;
        m_n = p.m_n;
        m_q = p.m_q;
        m_call = p.m_call;
    }
    return *this;
}
```

The computePriceStep member function is used to compute the immediate price for a single step of the algorithm. The indices i and j represent the position in the binomial tree. Other arguments are the necessary parameters used to calculate the price of this step. Notice that this member function is declared as virtual, and it can be later overridden for the use of American-style options.

```
void BinomialModel::computePriceStep(int i, int j, double K,
                                     vec &prices, double p_u, double p_d, double u)
{
    prices[i] = p_u * prices[i] + p_d * prices[i+1];
}
```

The main member function in the BinomialModel class is the function that computes the option price for a given strike, determined by the parameter K. The algorithm is essentially a C++ implementation of the ideas presented in the previous section. The first step is to calculate the price delta, using the period and the number of steps. Next the amount of price changes in the up side are calculated using the exp(m_sigma * sqrt(delta)) expression.

Next, the function computes the probabilities of moving up or down in the binomial tree, using the equations described previously. The probabilities are denoted by p_u and p_d.

```
double BinomialModel::optionPriceForStrike(double K)
{
    double delta = m_T / m_n;   // size of each step
    double u = exp(m_sigma * sqrt(delta));

    double p_u = (u * exp(-m_r * delta) - exp(-m_q * delta)) * u / (u*u - 1);
    double p_d =      exp(-m_r * delta) - p_u;

    vec prices(m_n);

    // compute last day values (leafs of the tree)
    for (int i= 0; i<m_n; ++i)
    {
        if (m_call)
        {
            prices[i] = std::max(0.0, m_S * pow(u, 2*i - m_n) - K);
        }
        else
        {
```

```
            prices[i] = std::max(0.0, K - m_S * pow(u, 2*i - m_n));
        }
    }

    for (int j = m_n-1; j>=0; --j)
    {
        for (int i = 0; i<j; ++i)
        {
            computePriceStep(i, j, K, prices, p_u, p_d, u);
        }
    }

    return prices[0];
}
```

The first `for` loop in this member function is responsible for computing the stock price at the last level of the binomial tree. This is done using the property that defines the price of an option at expiration. Therefore, there are two cases that need to be handled, depending on if the option is a call or a put.

The last `for` loop is the main computation that traverses the binomial tree from the last level to the root node. The step calculation is performed by the `computePriceStep` member function. The main idea, which you can see by looking at that member function, is to first compute the average (expected) price of the node. This is done by taking the expected value of the known prices that have been previously calculated according to the probabilities `p_u` and `p_d`.

After the option prices have been computed in this way, the algorithm will determine the price at the root node. Therefore, the price required is stored in position zero of the `prices` vector. The last line of this member function returns `prices[0]` as the desired solution.

---

■ **Note** The pricing strategy presented in this section works for options that cannot be exercised until the date of expiration. This type of option is commonly known as a European-style option. For American-style options, which can normally be exercised at any time, a slightly different pricing method needs to be used, as shown in the next section.

---

## Pricing American-Style Options

This section presents a slight modification of the binomial method that can be used to price American-style option contracts. An American-style option is defined in such a way that buyers of such options can exercise their rights (that is, buying or selling the underlying) at any time until expiration. This is in contrast to what is called European-style options, whereby option rights can be exercised only at expiration.

You can use the `AmericanBinomialModel` class to price American options. Looking at the code, you can see clearly how American options differ from European ones in terms of the option prices. The binomial model determines this by checking the possible exercise price of the option and taking that value into consideration if it is higher than the expected price.

The class interface is defined as follows. The public inheritance from `BinomialModel` allows you to share the methods defined in that class. The resulting interface is very simple because no additional member variables are necessary. It contains the standard copy constructor, a constructor that forwards the received parameters to the base class, and a destructor.

CHAPTER 12 ▪ BASIC MODELS FOR OPTIONS PRICING

```
#include <vector>
#include <cmath>

using vec = std::vector<double>;

class AmericanBinomialModel : public BinomialModel {
    AmericanBinomialModel(const BinomialModel &p);
    ~AmericanBinomialModel();
    AmericanBinomialModel &operator=(const BinomialModel &p);

    AmericanBinomialModel(double T, // expiration time
                double S,     // stock price
                double r,     // interest rate
                double sigma,
                double q,     // dividend yield
                int n,        // number of steps
                bool call
                );

    virtual void computePriceStep(int i, int j, double K, vec &prices,
                             double p_u, double p_d, double u);
};
```

The constructor just needs to forward the received parameters to the base class BinomialModel.

```
AmericanBinomialModel::AmericanBinomialModel(const BinomialModel &p)
: BinomialModel(p)
{
}

AmericanBinomialModel::~AmericanBinomialModel()
{
}
```

Because there are no extra member variables, the assignment operator can use the nice trick of calling the operator on the superclass to do the assignment work, as follows:

```
AmericanBinomialModel &AmericanBinomialModel::operator=(const BinomialModel &p)
{
    BinomialModel::operator=(p);  // no new data members in this class
    return *this;
}

AmericanBinomialModel::AmericanBinomialModel(double T, // expiration time
                      double S,     // stock price
                      double r,     // interest rate
                      double sigma,
                      double q,     // dividend yield
                      int n,        // number of steps
                      bool call)
: BinomialModel(T, S, r, sigma, q, n, call)
{
}
```

Next, you can see the real change that characterizes American options. The `computePriceStep` member function overrides the member function in the base class and allows the price of an American option to be calculated.

The first thing to do here is to call the member function from the superclass, so you don't need to repeat the same code, with potential duplication errors. Then, the function proceeds to calculate the exercise value. This is done by taking the adjusted stock price and subtracting it from the strike price. If the calculated exercise price is higher than the calculated price, then the price is updated with this exercise price. In other words, at each moment the price of the option has to be the highest of the potential value and the exercise value.

```
void AmericanBinomialModel::computePriceStep(int i, int j, double K, vec &prices, double p_u, double p_d, double u)
{
    BinomialModel::computePriceStep(i, j, K, prices, p_u, p_d, u);

    // compute exercise price for American option
    //
    double exercise = K - getStockPrice() * pow(u, 2*i - j);
    if (prices[i] < exercise)
    {
        prices[i] = exercise;
    }
}
```

## Solving the Black-Scholes Model

The previous sections explored discrete methods used to compute the price of options. These methods work by approximating the solution through the use of price trees, where each node represents a discrete step into the solution of the problem.

While the binomial tree method is appropriate in many situations, it is sometimes necessary to use a more rigorous method based on the Black-Scholes partial derivative equation (PDE). The model, develop by Black and Scholes in the 70s, provides a full mathematical description of how option prices evolved over time and with respect to the changes in the underlying prices.

The Black-Scholes model uses a few input parameters that describe the option and the conditions under which prices evolve. The parameters are:

- Expiration date
- Stock price
- Stock volatility
- Interest rates (paid on short term cash)
- Dividends paid by the underlying stock

Using these parameters, the model provides a partial derivative equation that contains the information necessary to determine the price of the option. The result from this model can be summarized in the following PDE:

$$\frac{\partial V}{\partial t} + \frac{1}{2}\sigma^2 S^2 \frac{\partial^2 V}{\partial S^2} + rS\frac{\partial V}{\partial S} = rV$$

In this differential equation, the quantities represented are:

- $V$: The price of the desired derivative
- $t$: The time
- $\sigma$: The volatility of the underlying stock
- $S$: The stock price
- $r$: The interest rate

If you know the previous information about the underlying security, such as prices, interest rates, and previous volatility, the Black-Scholes equation allows you to compute the value of a call or put option based on those assumptions. The solution of this equation can be achieved using several methods, such as simulation techniques and piecewise integration using numeric approximations. The next section presents a simple numeric technique that can be applied to find solutions to the Black-Scholes model.

## Numerical Solution of the Model

To solve the Black-Scholes model computationally, it is necessary to apply numerical techniques to solve the associated PDE. It is important to note that there are several methods used to compute this class of equations, with results that depend on the required accuracy, computational effort, and implementation difficulty.

This section explores a simple strategy to solve the Black-Scholes model. The strategy is based on what is called the *forward* method for the solution of PDEs. The forward method is an extension of Euler's method for the solution of ODEs, as described in the previous chapter. Unlike Euler's method, the forward method needs to find a solution for a differential equation that contains more than one variable.

The forward method solves this problem by dividing the domain of the desired equation into smaller, rectangular pieces, which can be easily computed. Once this is completed, the algorithm propagates those values forward, and at each step a small area $dS$ is considered.

For this method to work, it is necessary to provide a set of initial conditions for the PDE. In the case of options pricing, the natural set of initial conditions is the price at expiration, which is well known for each possible value of the stock. Therefore, the implementation of the forward in fact starts from the expiration date and proceeds backward in time to the desired date.

The C++ solution is implemented in the BlackScholesMethod class. This class provides a simple interface, where the main member function is called solve, and it returns the price at the desired date and under the conditions defined by the given parameters.

```
class BlackScholesMethod {
public:
    BlackScholesMethod(double expiration, double maxPrice, double strike, double intRate);
    BlackScholesMethod(const BlackScholesMethod &p);
    ~BlackScholesMethod();
    BlackScholesMethod &operator=(const BlackScholesMethod &p);

    std::vector<double> solve(double volatility, int nx, int timeSteps);
private:
    double m_expiration;
    double m_maxPrice;
    double m_strike;
    double m_intRate;
};
```

## CHAPTER 12 ■ BASIC MODELS FOR OPTIONS PRICING

In the implementation file, which is listed next, you will first find the constructors and assignment operator. These member functions just initialize the private variables, which include:

- Expiration date, denoted by m_expiration
- Maximum price that will be considered by the algorithm, denoted by m_maxPrice
- Strike price, denoted by m_strike
- Current interest rate, denoted by m_intRate

```
#include "BlackScholes.hpp"

#include <cmath>
#include <algorithm>
#include <vector>
#include <iostream>
#include <iomanip>

using std::vector;
using std::cout;
using std::endl;
using std::setw;

BlackScholesMethod::BlackScholesMethod(double expiration, double maxPrice,
                                       double strike, double intRate)
: m_expiration(expiration),
m_maxPrice(maxPrice),
m_strike(strike),
m_intRate(intRate)
{
}

BlackScholesMethod::BlackScholesMethod(const BlackScholesMethod &p)
: m_expiration(p.m_expiration),
m_maxPrice(p.m_maxPrice),
m_strike(p.m_strike),
m_intRate(p.m_intRate)
{
}

BlackScholesMethod::~BlackScholesMethod()
{
}

BlackScholesMethod &BlackScholesMethod::operator=(const BlackScholesMethod &p)
{
    if (this != &p)
    {
        m_expiration = p.m_expiration;
        m_maxPrice = p.m_maxPrice;
        m_strike = p.m_strike;
```

## CHAPTER 12 ▪ BASIC MODELS FOR OPTIONS PRICING

```
        m_intRate = p.m_intRate;
    }
    return *this;
}
```

The solve method is the heart of the algorithm. The first part of this member function is responsible for initializing common expressions that are used throughout the algorithm. These expressions are stored in vectors a, b, and c. In mathematical notation, these factors can be presented as:

$$a_n = \frac{1}{2}\left(nrdt - (nV)^2 dt\right)$$

$$b_n = 1 - rdt + (nV)^2 dt$$

$$c_n = \frac{1}{2}\left(nrdt + (nV)^2 dt\right)$$

The third for loop is the place where the initial conditions are prepared, by direct calculation of the price at expiration date. The last loop is where the forward algorithm is used. Each step of the loop will compute the contributions for that particular time period, assuming that the period j-1 is known. At the end the u vector, where the option prices have been stored, is returned to the caller.

```
vector<double> BlackScholesMethod::solve(double volatility, int nx, int timeSteps)
{
    double dt = m_expiration /(double)timeSteps;
    double dx = m_maxPrice /(double)nx;

    vector<double> a(nx-1);
    vector<double> b(nx-1);
    vector<double> c(nx-1);

    int i;
    for (i = 0; i < nx - 1; i++)
    {
        b[i] = 1.0 - m_intRate * dt - dt * pow(volatility * (i+1), 2);
    }

    for (i = 0; i < nx - 2; i++)
    {
        c[i] = 0.5 * dt * pow(volatility * (i+1), 2) + 0.5 * dt * m_intRate * (i+1);
    }

    for (i = 1; i < nx - 1; i++)
    {
        a[i] = 0.5 * dt * pow(volatility * (i+1), 2) - 0.5 * dt * m_intRate * (i+1);
    }

    vector<double> u((nx-1)*(timeSteps+1));

    double u0 = 0.0;
```

```cpp
    for (i = 0; i < nx - 1; i++)
    {
        u0 += dx;
        u[i+0*(nx-1)] = std::max(u0 - m_strike, 0.0);
    }

    for (int j = 0; j < timeSteps; j++)
    {
        double t = (double)(j) * m_expiration /(double)timeSteps;

        double p = 0.5 * dt * (nx - 1) * (volatility*volatility * (nx-1) + m_intRate)
            * (m_maxPrice-m_strike * exp(-m_intRate*t ) );

        for (i = 0; i < nx - 1; i++)
        {
            u[i+(j+1)*(nx-1)] = b[i] * u[i+j*(nx-1)];
        }
        for (i = 0; i < nx - 2; i++)
        {
            u[i+(j+1)*(nx-1)] += c[i] * u[i+1+j*(nx-1)];
        }
        for (i = 1; i < nx - 1; i++)
        {
            u[i+(j+1)*(nx-1)] += a[i] * u[i-1+j*(nx-1)];
        }
        u[nx-2+(j+1)*(nx-1)] += p;
    }

    return u;
}
```

Finally, I present a simple test function that can be used to illustrate the use of the BlackScholesMethod class. This function first initializes some parameters with reasonable values. Then, it creates a new object of type BlackScholesMethod, passing to the constructor some of the previously defined parameters.

The blackSholes object is then used to solve the pricing problem. The result is a vector of prices, one for each of the steps used by the algorithm (in practice, only the last value would be used). Finally, the function prints the result so that you can inspect the convergence of the algorithm.

```cpp
void test_bsmethod()
{
    auto strike = 5.0;
    auto intRate = 0.03;
    auto sigma = 0.50;
    auto t1 = 1.0;
    auto numSteps = 11;
    auto numDays = 29;
    auto maxPrice = 10.0;

    BlackScholesMethod blackScholes(t1, maxPrice, strike, intRate);
    vector<double> u = blackScholes.solve(sigma, numSteps, numDays);
```

```
        double minPrice = .0;
        for (int  i=0; i < numSteps-1; i++)
        {
            double s = ((numSteps-i-2) * minPrice+(i+1)*maxPrice)/ (double)(numSteps-1);
            cout << "  " << s << "  " << u[i+numDays*(numSteps-1)] << endl;
        }
    }
}
```

## Complete Code

This section presents the complete code for the BlackScholesMethod class. The code depends only on the STL and functions in the standard C++ library. As such, it can serve as a first step toward a complete solution for options valuation processes.

The code is divided into a header file called BlackScholes.hpp and an associated implementation file. These files are presented in Listings 12-1 and 12-2, respectively.

*Listing 12-1.* Header File for the BlackScholesMethod Class

```
//
//  BlackScholes.hpp

#ifndef BlackScholes_hpp
#define BlackScholes_hpp

#include <vector>

class BlackScholesMethod {
public:
    BlackScholesMethod(double expiration, double maxPrice, double strike, double intRate);
    BlackScholesMethod(const BlackScholesMethod &p);
    ~BlackScholesMethod();
    BlackScholesMethod &operator=(const BlackScholesMethod &p);
    
    std::vector<double> solve(double volatility, int nx, int timeSteps);
private:
    double m_expiration;
    double m_maxPrice;
    double m_strike;
    double m_intRate;
};
#endif /* BlackScholes_hpp */
```

*Listing 12-2.* Implementation File for the BlackScholesMethod Class

```
//
//  BlackScholes.cpp

#include "BlackScholes.hpp"

#include <cmath>
#include <algorithm>
```

CHAPTER 12 ■ BASIC MODELS FOR OPTIONS PRICING

```
#include <vector>
#include <iostream>
#include <iomanip>

using std::vector;
using std::cout;
using std::endl;
using std::setw;

BlackScholesMethod::BlackScholesMethod(double expiration, double maxPrice,
                                       double strike, double intRate)
: m_expiration(expiration),
m_maxPrice(maxPrice),
m_strike(strike),
m_intRate(intRate)
{
}

BlackScholesMethod::BlackScholesMethod(const BlackScholesMethod &p)
: m_expiration(p.m_expiration),
m_maxPrice(p.m_maxPrice),
m_strike(p.m_strike),
m_intRate(p.m_intRate)
{
}

BlackScholesMethod::~BlackScholesMethod()
{
}

BlackScholesMethod &BlackScholesMethod::operator=(const BlackScholesMethod &p)
{
    if (this != &p)
    {
        m_expiration = p.m_expiration;
        m_maxPrice = p.m_maxPrice;
        m_strike = p.m_strike;
        m_intRate = p.m_intRate;
    }
    return *this;
}

vector<double> BlackScholesMethod::solve(double volatility, int nx, int timeSteps)
{
    double dt = m_expiration /(double)timeSteps;
    double dx = m_maxPrice /(double)nx;

    vector<double> a(nx-1);
    vector<double> b(nx-1);
    vector<double> c(nx-1);
```

203

```cpp
    int i;
    for (i = 0; i < nx - 1; i++)
    {
        b[i] = 1.0 - m_intRate * dt - dt * pow(volatility * (i+1), 2);
    }

    for (i = 0; i < nx - 2; i++)
    {
        c[i] = 0.5 * dt * pow(volatility * (i+1), 2) + 0.5 * dt * m_intRate * (i+1);
    }

    for (i = 1; i < nx - 1; i++)
    {
        a[i] = 0.5 * dt * pow(volatility * (i+1), 2) - 0.5 * dt * m_intRate * (i+1);
    }

    vector<double> u((nx-1)*(timeSteps+1));

    double u0 = 0.0;
    for (i = 0; i < nx - 1; i++)
    {
        u0 += dx;
        u[i+0*(nx-1)] = std::max(u0 - m_strike, 0.0);
    }

    for (int j = 0; j < timeSteps; j++)
    {
        double t = (double)(j) * m_expiration /(double)timeSteps;

        double p = 0.5 * dt * (nx - 1) * (volatility*volatility * (nx-1) + m_intRate)
        * (m_maxPrice-m_strike * exp(-m_intRate*t ) );

        for (i = 0; i < nx - 1; i++)
        {
            u[i+(j+1)*(nx-1)] = b[i] * u[i+j*(nx-1)];
        }
        for (i = 0; i < nx - 2; i++)
        {
            u[i+(j+1)*(nx-1)] += c[i] * u[i+1+j*(nx-1)];
        }
        for (i = 1; i < nx - 1; i++)
        {
            u[i+(j+1)*(nx-1)] += a[i] * u[i-1+j*(nx-1)];
        }
        u[nx-2+(j+1)*(nx-1)] += p;
    }

    return u;
}

int main()
{
```

```
        auto strike = 5.0;
        auto intRate = 0.03;
        auto sigma = 0.50;
        auto t1 = 1.0;
        auto numSteps = 11;
        auto numDays = 29;
        auto maxPrice = 10.0;

        BlackScholesMethod blackScholes(t1, maxPrice, strike, intRate);
        vector<double> u = blackScholes.solve(sigma, numSteps, numDays);

        double minPrice = .0;
        for (int  i=0; i < numSteps-1; i++)
        {
            double s = ((numSteps-i-2) * minPrice+(i+1)*maxPrice)/ (double)(numSteps-1);
            cout << "  " << s << "  " << u[i+numDays*(numSteps-1)] << endl;
        }
        return 0;
}
```

# Conclusion

Options pricing is a very common problem that needs to be solved if you need to trade these types of financial derivatives. Because underlying prices change so frequently, it is very important that option prices be calculated efficiently. C++ is an ideal language for encoding the solution to these pricing problems.

In this chapter, you saw an introduction to the most common strategies for options pricing. The most popular techniques can be divided into lattice models, such as binomial trees, and PDE-based algorithms, where the Black-Scholes model or some close variation is solved through the use of numerical methods for PDEs.

The first sections of this chapter demonstrated the binomial method, with its assumptions and mathematical ideas. You learned how these ideas can be used in C++ and encapsulated into a class. The model was extended to deal with American-style options, where option buys have the ability to exercise the option at any time before the (or at the) expiration date.

You also saw how to represent the options pricing problem in terms of the Black-Scholes model, which uses a PDE that describes the changes in options pricing. This model is solved using a method that discretizes the domain of the function and calculates the result in a large number of small steps.

In the next chapter, you will learn about Monte Carlo methods, another strategy that is commonly used to solve problems in the area of mathematical finance. In particular, Monte Carlo methods can be used to efficiently solve some difficult problems of derivative pricing without needing to directly compute probabilities, as used by the methods discussed in this chapter.

# CHAPTER 13

# Monte Carlo Methods

Among other programming techniques used for trading equity markets, the Monte Carlo simulation has a special place due to factors such as its wide applicability and easy implementation. These methods can be used to implement strategies for market analysis such as price forecasting, or to validate options trading strategies, for example.

A great advantage of the Monte Carlo methods is the fact that they can be used to study complex events without the need to solve complicated mathematical models and equations. Using the idea of simulation through the use of random numbers, Monte Carlo methods offer the ability to study a large class of events, which would otherwise be difficult to analyze using exact techniques.

This chapter provides an introduction to stochastic methods and how they be used as part of simulation-based algorithms applied to options pricing. Here are a few of the topics that will be covered in this chapter:

- *Random number generation*: Generating random numbers is a basic step in creating algorithms that exploit stochastic behavior. Monte Carlo methods require the use of effective random number generation routines, which will be discussed in this chapter.

- *Probability distributions*: Monte Carlo algorithms are based on the properties of stochastic events. Many of these events occur according to well-known probability distributions. In C++, it is possible to generate numbers according to many popular probability distributions, as you will learn.

- *Random walks*: A random walk is a stochastic process where a certain quantity can randomly change with equal probability to positive or negative side. This makes random walk very useful for modeling prices in financial markets, as well as for simulating trading strategies.

- *Stochastic models for options pricing*: Another application of random walks is in the determination of option prices. Using a stochastic method for this purpose is useful if you want to avoid the use of a more complex exact or approximate model, such as the algorithms described in the previous chapter.

## Introduction to Monte Carlo Methods

A Monte Carlo algorithm is a computational procedure that uses random numbers to simulate and study complex events. It is based on the idea that you can analyze the results of an event by repeating it several times in different ways, with the help of a computer or other method to generate random numbers.

This idea behind Monte Carlo methods is not new, having been used for as long as probability methods have been studied. For example, a well-known randomized procedure to determine the area of a geometric shape is to throw darts at the figure. After a while, you can count the percentage of darts inside the shape and use that percentage to determine the area.

CHAPTER 13 ■ MONTE CARLO METHODS

Despite their simplicity, Monte Carlo methods may be time consuming, and they require a large number of repetitions to achieve their goals. The recent development of fast computers made possible to use such methods in an increasing number of situations, making them practical and capable of finding solutions for problems where explicit mathematical analysis is very difficult.

In general, Monte Carlo methods have been used for the solution of mathematical and computational problems where it is difficult to perform direct observations. Algorithms based on Monte Carlo methods use simulation strategies to determine values that normally occur as the result of random events in several areas, including the financial markets. In fact, the application of Monte Carlo to finance methods is widespread. You will find many algorithms used in the analysis of options and derivatives that exploit Monte Carlo techniques. For example:

- *Options pricing*: It is possible to use randomized algorithms to determine the prices of options and other derivatives.

- *Trade strategy analysis*: Monte Carlo methods can be used to test different trade strategies using simulated prices. This type of analysis is invaluable, since it allows you to test trading techniques on a large amount of data that is independent of the existing market observations.

- *Analysis of bonds and other fixed income investments*: Bonds and their derivatives are tied to fluctuations of interest rates over different time horizons. An effective way to study the behavior of bonds is to construct stochastic models and use them to perform an analysis.

- *Portfolio analysis*: Another area where Monte Carlo methods are useful is when studying a portfolio of investments. The stochastic algorithm allows analysts to vary the rate of exposure to diverse economic scenarios and try to determine the best allocation for a portfolio.

In the next few sections, you will first learn the tools necessary to design and implement Monte Carlo algorithms using the C++ language. You will also see examples of how these tools can be used to analyze options and related instruments.

## Random Number Generation

The first topic that is addressed is random number generation. True random numbers are not possible to achieve in digital computers, but there are several techniques to create sequences of pseudo-random numbers. These methods have been made available through the standard C++ libraries, as will be covered in this section.

For C++ programmers, the main source of random number generation routines is the `<random>` header file provided by the standard library. With these functions, you can generate pseudo-random numbers that are well tested and that can be accessed through an easy interface.

The first thing to learn about random number generation in the standard library is the concept of generators. A *generator* can be viewed as a source of pseudo-random bits, that is, an algorithm that is capable of returning numbers that are uniformly random. The C++ library offers a small number of generators that can be used by programmers. Here are some of the available generators:

- *Mersenne twister*: This is one of the most popular generators. It is based on an algorithm that uses Mersenne prime numbers as the period length of the sequence of pseudo-random numbers. The Mersenne twister algorithm is considered to be one of the best general-purpose generators of random numbers and it is frequently used in applications.

- *Linear congruential engine*: This engine is based on a traditional algorithm that uses simple addition, multiplication, and module operations to produce numbers that have pseudo-random properties. This generator is indicated when you need fast sequences of random numbers, due to its efficiency. However, the linear congruential algorithm is known to generate numbers that possess some correlation.
- *Subtract with carry*: This is still another algorithm that is used to generate random numbers in the standard library. The algorithm is called lagged Fibonacci, and it uses a numeric sequence that has properties that are similar to the famous Fibonacci sequence.

These generators represent three of the most common ways to generate random numbers. Other techniques for random number generation have also been proposed in the scientific literature. Table 13-1 shows some of the most commonly used algorithms for random number generation.

*Table 13-1. Algorithm for Pseudo-Random Number Generation*

| Algorithm | Description |
| --- | --- |
| Linear congruential | Traditional method that uses modulo arithmetic. |
| Inversive congruential | Uses the modular multiplicative inverse to generate new elements in the sequence. |
| Mersenne twister | Method developed in 1997; uses Mersenne primes to generate random numbers. |
| WELL generators | Well Equidistributed Long-Period Linear, based on the application for operations on a binary field. |
| XorShift generators | Fast method that uses exclusive-or operations to generate new random numbers. |
| Linear feedback shift | Method that uses a linear function over the existing sequence of values to generate the next random number. |
| Park-Miller generator | A linear congruential generator that uses multiplicative groups of integers under the modulo operation. |

The second part of the random generation library in C++ is the use of engine instantiations. These instantiations can be viewed as a concrete implementation of a generic algorithm. For example, consider the Mersenne twister engine, which is implemented as a template called mersenne_twister_engine., The easiest way to use this engine is to apply an instantiation such as the minstd_rand (minimal standard pseudo-random number) generator. This particular instantiation is defined by the C++ standard as:

```
typedef linear_congruential_engine<
        uint_fast32_t,
        48271,
        0,
        2147483647> minstd_rand;
```

The linear_congruential_engine is a common random generator engine that is implemented by the standard library. A list of known engine instantiations in the C++ standard library are presented in Table 13-2. You can choose one of these instantiations as a generator for your own algorithm, or you can create a new instantiation.

*Table 13-2. A List of Generator Instantiations Available on the Standard Library*

| Generator Instantiation | Parameters |
|---|---|
| default_random_engine | Random engine that is provided as a default option by the library implementation. |
| knuth_b | Defined as typedef shuffle_order_engine <minstd_rand0,256> knuth_b;. |
| minstd_rand | Minimal standard generator; it is an instantiation of linear_congruential_engine. |
| minstd_rand0 | Similar to the engine described above, with particular parameters. |
| mt19937 | Mersenne twister generator. |
| mt19937_64 | Mersenne twister generator for 64-bit types. |
| ranlux24 | Uses the subtract-with-carry generator and returns values that use a 24-bit representation. |

■ **Note** Random number generators can be freely instantiated in the standard library. However, you should rarely need to define a new instantiation, unless you have good knowledge about how the parameters for each generator work together. A careful study of parameters is usually necessary to create a new generator, since they are based on statistical properties that have been determined after careful analysis made by researchers in the area.

The generators and their instantiations can be thought of as the original source for pseudo-random bits. Once you have defined a source, it is possible to generate random numbers according to a given probability distribution, as you will see in the next section.

## Probability Distributions

A probability distribution is family of functions that defines the parameters for a stochastic process. For example, the simplest distribution of random numbers is the uniform distribution, where each value is generated with equal probability in a given range. A particular case of the uniform distribution is Uniform[0,1], where each number is randomly generated with equal probability in the range between 0 and 1.

There are a small number of probability distributions that occur very frequently in the analysis of natural events. These common distributions, which have been studied in several branches of stochastic analysis, are now available as part of the C++ <random> header in standard library. For examples of two common probability distributions, see Figure 13-1 (which shows the Normal distribution with mean 0 and standard deviation 1) and Figure 13-2 (which shows the Exponential distribution with mean 1).

CHAPTER 13 ■ MONTE CARLO METHODS

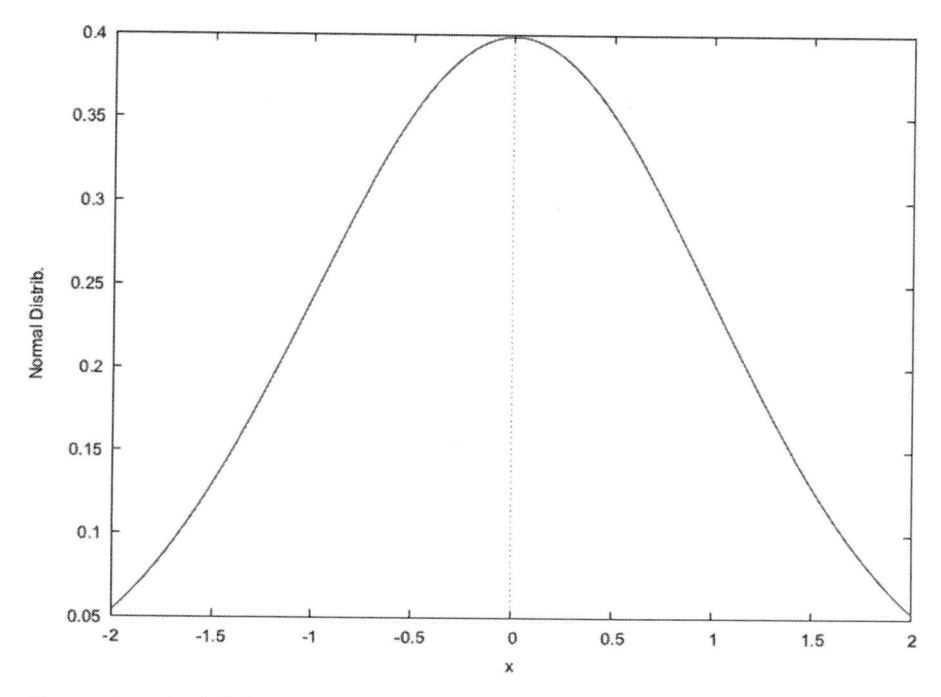

*Figure 13-1.* *Probabilities defined by the Normal distribution, with mean 0 and standard deviation 1*

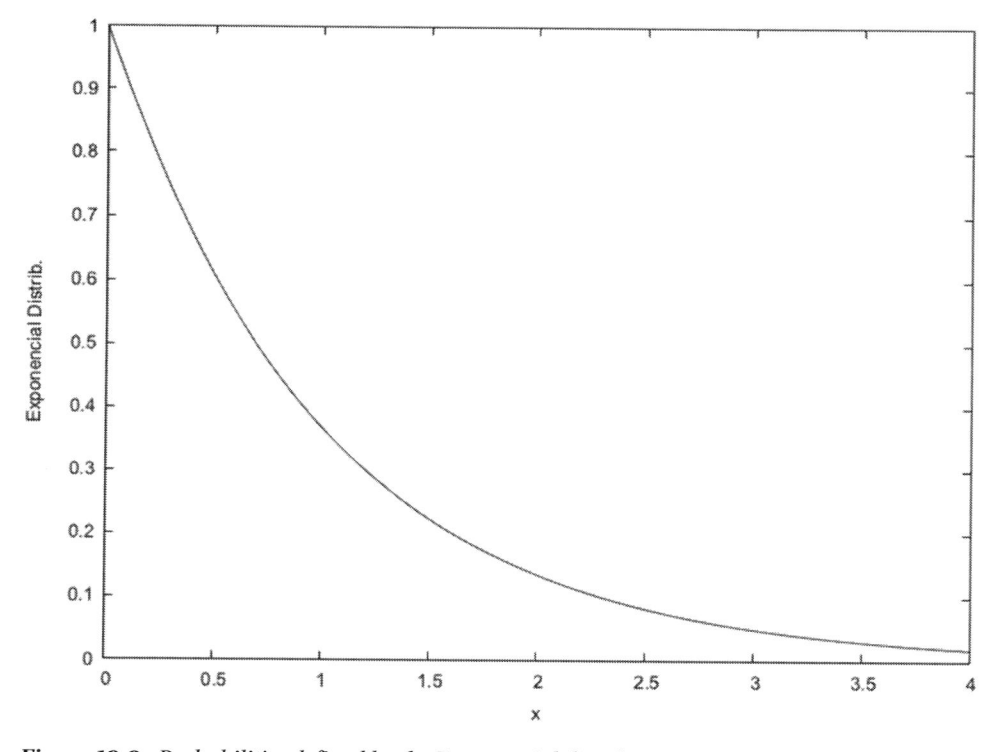

*Figure 13-2.* *Probabilities defined by the Exponential distribution, with mean 1*

CHAPTER 13 ■ MONTE CARLO METHODS

Consider the most common case of generating uniform random integer numbers in a particular range. This can be easily handled in the standard library by using the std::uniform_int_distribution template. This template is capable of creating integer numbers that have uniform distribution as given by the two parameters: the initial and maximum values. Here is an example of how to code a function that returns such random integer numbers.

```
#include <iostream>
#include <random>

using std::cout;
using std::endl;

std::default_random_engine generator;

int get_uniform_int(int max)
{
    if (max < 1)
    {
        cout << "invalid parameter max " << max << endl;
        throw std::runtime_error("invalid parameter max");
    }
    std::uniform_int_distribution<int> uint(0,max);

    return uint(generator);
}
```

The first step is to define a generator to use as the source of random bits. This is done by instantiating an engine (done at the file scope). The std::default_random_engine is the default generator selected by the compiler's implementation. It should be a reasonable choice, unless you want to be very specific about the generator for your code.

The get_uniform_int function generates a random integer between 0 and max, where max is a parameter passed to the function. The function first checks if the parameter is valid and throws an exception when that is not the case. The function then uses the parameter to create an object of type uniform_int_distribution. This object receives two parameters that define the distribution: the minimum and maximum values. The resulting object is then used to generate the random number itself.

■ **Note** Traditional C and C++ code used to rely on the rand function to generate random integer numbers. This usage is now deprecated because the algorithm used in rand() is known to have weaknesses. In particular, the idea of using the expression (rand() % N) to generate random integer numbers in the range 0 to N-1 has been proved to be unreliable. Even though the numbers seem random enough for most applications, it fails when you try to perform more complex statistical analysis.

The sequence of steps to use the random number generators and distributions are therefore summarized as follows:

- Find a suitable random engine and a corresponding generator according to the needs of your application.

- Select a generator instantiation based on the random engine you selected previously. If you don't have any specific requirements, the default_random_engine could be used.

- Select a random distribution according to the needs of your application. A common distribution is the uniform, which produces numbers with the same probability in a given range.

- Create an object of the type determined by the probability distribution. In the previous example, you used uniform_int_distribution as the object type.

- The resulting object can now be called to generate pseudo-random numbers, once the generator object is passed as the single parameter for the call. This makes it possible to use generators of different types or, more commonly, generators that are used for a specific function of a thread.

## Using Common Probability Distributions

This section will show a few examples of common probability distributions and how they can be used in C++. As mentioned, random numbers can be generated according to different probability functions. These families of functions are grouped according to the parameters and shape of the distribution.

One of the simplest probability distributions is the Bernoulli distribution. This is a family of probability distributions that model a yes/no scenario, an event that has only two results. The only parameter for this distribution is the probability of the yes result. The simplest example of this type of model is a coin toss, with parameter 0.5, representing a fair probability of heads or tails.

In the next code example, the function coin_toss_experiment returns a vector of Boolean values, representing the result of a set of fair coin tosses.

```
#include <iostream>
#include <random>
#include <vector>

using std::cout;
using std::endl;
using std::vector;

std::default_random_engine generator;

vector<bool> coin_toss_experiment(int num_experiments)
{
    if (num_experiments < 1)
    {
        cout << "invalid number of experiments " << num_experiments << endl;
        throw std::runtime_error("invalid number of experiments");
    }

    std::bernoulli_distribution bernoulli(0.5);

    vector<bool> results;
    for (int i=0; i<num_experiments; ++i)
    {
        results.push_back(bernoulli(generator));
    }
    return results;
}
```

## CHAPTER 13 ▪ MONTE CARLO METHODS

In this code, the first step is to use a generator, which in this case is std::default_random_engine allocated in the file scope, so it is available during the lifetime of the application. The coin_toss_experiment function initially checks the validity of the parameter num_experiments, which gives the number of tries in this random experiment.

The function then allocates a new object from the Bernoulli distribution, with parameter 0.5, which indicates that the yes/no event occurs with even probability for each side. The random values are then generated in the loop, where the bernoulli returns Boolean values according to the desired distribution behavior. The values are stored in a vector<bool> container.

Another common distribution that is used to model natural events is the *Poisson distribution*. This distribution arises commonly when observing the number of events that occur in a period of time, under the assumption that these events are independent. For example, the number of customers arriving at a coffee shop during a given period could be modeled as a Poisson distribution. The mathematical expression used to model the probability distribution of such events is given by:

$$p(k) = \frac{\lambda^k e^{-k}}{k!}$$

Here, $k$ is the number events that are observed, and $\lambda$ is the parameter that determines the results of the experiment, which can be interpreted as the average number of events occurring in the given time period.

In the C++ standard library, the Poisson distribution is made available through the std::poisson_distribution template. The parameter for this distribution is the mean, usually represented as the mathematical variable $\lambda$ as in the previous equation.

The following is an example that can be used to analyze the number of customers buying in a particular store in a time period. For instance, financial analysts perform this type of study when they need to study the buying patterns at a particular business. The code defines a function named num_customers_experiment:

```
#include <iostream>
#include <random>
#include <vector>

using std::cout;
using std::endl;
using std::vector;

vector<int> num_customers_experiment(double mean, int max, int ntries)
{
    std::default_random_engine generator;

    vector<int> occurrences(max, 0);
    std::poisson_distribution<int> poisson(mean);

    for (int i=0; i<ntries; ++i)
    {
        int result = poisson(generator);
        if (result < max) {
            occurrences[result] ++;
        }
    }

    return occurrences;
}
```

CHAPTER 13 ■ MONTE CARLO METHODS

The num_customers_experiment function can generate a sequence of random values based on the Poisson distribution and return a histogram of these values, that is, for each value it returns the number of times this value was observed.

The algorithm is similar to what you have seen before with the Bernoulli distribution. The first part is used to define the random generator, and it creates an object of type std::poisson_distribution. The parameter passed represents the mean of the distribution.

The for loop in the algorithm is used to build the histogram. At each step, a number is generated according to the Poisson distribution. Then, if the resulting number is less than the parameter max, that value is incremented in the list of occurrences.

The num_customers_experiment function is used in the next code fragment to print the results of the calculation. These numbers have been saved and used to create the chart displayed in Figure 13-3, which shows the observations between 0 and 20 and the corresponding number of observations for 200 trials.

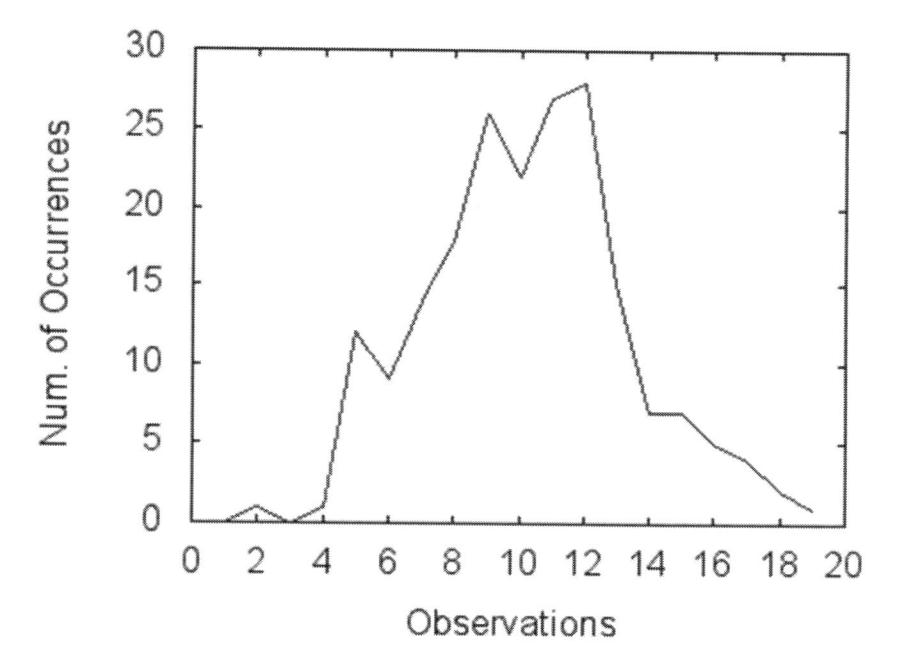

*Figure 13-3. Histogram of the data returned by function num_customers_experiment*

```
int test_experiment()
{
    auto data = num_customers_experiment(10.5, 20, 200);

    for (int i=0; i<int(data.size()); ++i)
    {
        cout << " event " << i << " occurred "  << data[i] << " times" << endl;
    }
}
```

The next example shows how to generate and use random values drawn from the normal distribution. The normal distribution, also known as Gaussian distribution, is one of the most common probability distributions used to model real world data. It is employed in data analysis, in areas ranging from drug

215

## CHAPTER 13 ■ MONTE CARLO METHODS

design to sociology. The normal distribution represents the distribution of values that are naturally measured in populations. For example, the heights of people living in a particular geographical area follow the normal distribution.

The bell-shaped probability graph of the normal distribution is determined by the Gaussian equation, which takes as parameters the mean and the standard deviation of a random variable. The equation is given by:

$$p(x) = \frac{1}{\sigma\sqrt{2\pi}} \exp\left(-\frac{(x-\mu)^2}{2\sigma^2}\right)$$

In this equation, $\mu$ is the mean value of these numbers, and $\sigma$ is the standard deviation, which is a measure of the variability of these random values.

In the following code example, you will see how to generate numbers that follow the normal distribution. The get_normal_observations function returns a list of numbers that have been generated according to the normal distribution according to the parameters mean and stdev.

```
#include <iostream>
#include <random>
#include <vector>
#include <assert.h>

using std::cout;
using std::endl;
using std::vector;

vector<double> get_normal_observations(int n, double mean, double stdev)
{
    std::default_random_engine generator;

    vector<double> values;
    std::normal_distribution<double> normaldist(mean, stdev);

    for (int i=0; i<n; ++i)
    {
        values.push_back(normaldist(generator));
    }

    return values;
}
```

The next function, test_normal, can be used to verify the correctness of this code. The idea of this function is to use the generated values so that it can create a histogram of the normal-distributed data. The first step of the algorithm is to call the get_normal_observations function and save the returned data. The next step is to get some information about the received data, such as the minimum and maximum values. This is done using the std::minmax_element function, which returns a pair of iterators pointing to the minimum and maximum values in the given range.

The algorithm creates a vector with elements corresponding to "bins," that is, smaller ranges where each observation is recorded. The size of each such bin is stored as the variable h. The first loop then determines the number of elements in each such range so that a histogram can be calculated.

The second loop is responsible for printing the results of the histogram. Each value is printed along with the starting point of the corresponding range.

# CHAPTER 13 ■ MONTE CARLO METHODS

```
void test_normal()
{
    vector<double> nv = get_normal_observations(1000, 8, 2);

    auto res = std::minmax_element(nv.begin(), nv.end());
    double min = *(res.first);
    double max = *(res.second);

    int N = 100;
    double h = (max - min)/double(N);
    vector<int> values(N, 0);

    for (int i=0; i<int(nv.size()); ++i)
    {
        double v = nv[i];
        int pos = int((v - min) / h);
        if (pos == N) pos--; // avoid the highest element
        values[pos]++;
    }
    for (int i=0; i<N; ++i)
    {
        cout << min + (i*h) << " " << values[i] << endl;
    }
}
```

The values created in this way have been plotted and are displayed in Figure 13-4. The horizontal axis represents the value of each observation. The vertical axis represents the number of occurrences of each observation.

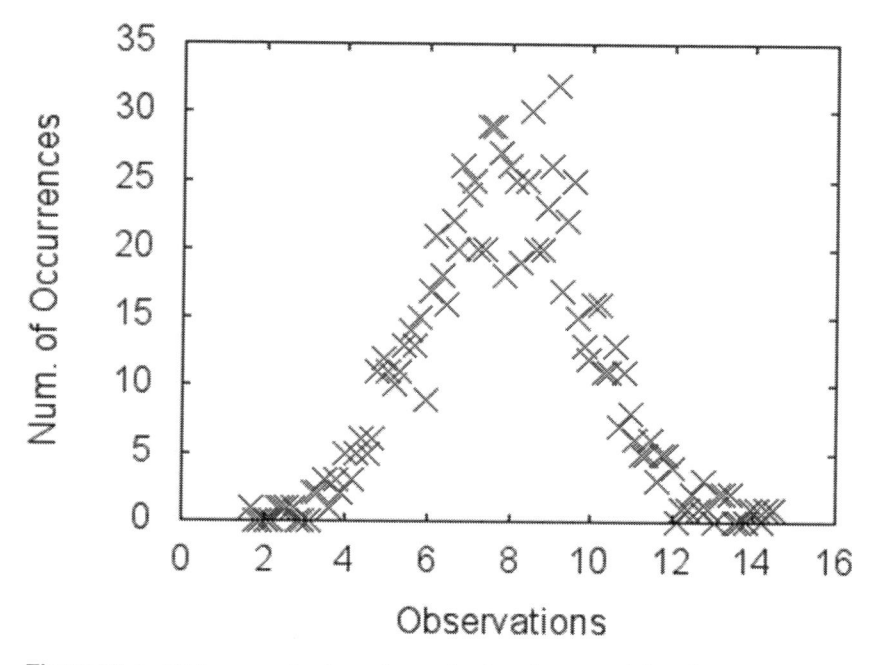

*Figure 13-4. Histogram of values observed using the normal distribution with mean 8*

CHAPTER 13 ■ MONTE CARLO METHODS

# Creating Random Walks

One of the main applications of stochastic processes in finance is the study of prices under random variations. This random process is called a random walk, since it implies that changes happen at random as time passes. A random walk model can be used to simulate market conditions and investigate the behavior of trades strategies, portfolios, and market participants in general. In this section, you see how to create a simple random walk using some of the facilities provided by C++.

A random walk can be designed with the use of a few simple rules that determine the price fluctuations. Notice the exact rules used depend on the kind of market that you need to simulate and the exact conditions that need to be replicated. In this example, I use a few computational commands that will simplify the task; the framework can be readily extended to implement more complex scenarios.

The random walk starts at an initial price given as a parameter to the algorithm. At each step, there are three possibilities for the random walk:

- A price decrease, which occurs with probability 1/3.
- A price increase, also happening with probability 1/3.
- The price remains unchanged.

The amount of increase or decrease is given by a parameter called stepSize.

These rules are implemented in the RandomWorkModel class. The class has an interface that exposes two member functions. getWalk returns a vector with a set of steps in the random walk.

```
//
//  RandomWalk.hpp

#ifndef RandomWalk_hpp
#define RandomWalk_hpp

#include <vector>

// Simple random walk for price simulation
class RandomWalkModel {
public:
    RandomWalkModel(int size, double start, double step);
    RandomWalkModel(const RandomWalkModel &p);
    ~RandomWalkModel();
    RandomWalkModel &operator=(const RandomWalkModel &p);

    std::vector<double> getWalk();
private:
    int random_integer(int max);

    int m_numSteps;        // number of steps
    double m_stepSize;     // size of each step (in percentage)
    double m_startPrice;   // starting price
};

#endif /* defined(__FinancialSamples__RandomWalk__) */
```

## CHAPTER 13 ■ MONTE CARLO METHODS

The class interface also contains the following member variables:

- The number of steps, m_numSteps, determines the number of steps (time) in the random walk.
- The initial price is defined by the m_stepSize member variable.
- The starting price is defined by the m_startPrice member variable.

These member variables are initialized in the constructor of RandomWalkModel, as shown in this code listing:

```
//
//   RandomWalk.cpp

#include "RandomWalk.hpp"

#include <cstdlib>
#include <iostream>
#include <random>

using std::vector;
using std::cout;
using std::endl;

std::default_random_engine engine;

RandomWalkModel::RandomWalkModel(int size, double start, double step)
: m_numSteps(size),
  m_stepSize(step),
  m_startPrice(start)
{
}

RandomWalkModel::RandomWalkModel(const RandomWalkModel &p)
: m_numSteps(p.m_numSteps),
  m_stepSize(p.m_stepSize),
  m_startPrice(p.m_startPrice)
{
}

RandomWalkModel::~RandomWalkModel()
{
}

RandomWalkModel &RandomWalkModel::operator=(const RandomWalkModel &p)
{
    if (this != &p)
    {
        m_numSteps = p.m_numSteps;
        m_stepSize = p.m_stepSize;
        m_startPrice = p.m_startPrice;
    }
    return *this;
}
```

219

CHAPTER 13 ■ MONTE CARLO METHODS

The random numbers needed by this code are generated using the random_integer member function. This function just uses the standard library random number generator std::default_random_engine. It also uses the uniform distribution returning integer values, as provided by the std::uniform_distribution template class.

```
int RandomWalkModel::random_integer(int max)
{
    std::uniform_int_distribution<int> unif(0, max);
    return unif(engine);
}
```

The random walk sequence is generated by the member function getWalk. The algorithm has a single loop that repeats the price generation according to the m_numSteps variable. Inside the loop, the code selects a random integer between zero and 2. Depending on the result, the code makes a decision to increase, decrease, or leave the price unchanged. Each price is then added to a vector, and the vector is returned at the end of the function.

```
std::vector<double> RandomWalkModel::getWalk()
{
    vector<double> walk;
    double prev = m_startPrice;

    for (int i=0; i<m_numSteps; ++i)
    {
        int r = random_integer(3);
        cout << r << endl;
        double val = prev;
        if (r == 0) val += (m_stepSize * val);
        else if (r == 1) val -= (m_stepSize * val);
        walk.push_back(val);
        prev = val;
    }
    return walk;
}
```

This code can be tested using the test_random_walk function. This function simply creates a RandomWalkModel object with 200 steps, starting at the $30 price and with steps of $0.01.

```
int test_random_walk()
{
    RandomWalkModel rw(200, 30, 0.01);
    vector<double> walk = rw.getWalk();
    for (int i=0; i<walk.size(); ++i)
    {
        cout << ", " << walk[i];
    }
    cout << endl;
    return 0;
}
```

The random walk generated by the test_random_walk function was saved, and using that data I plotted the results, as shown in Figure 13-5. Notice that, although this model is very simple, the results are not very different from what is observed in the market. Using this kind of synthetic data, you can test trading strategies and determine if they are profitable in such randomized scenarios.

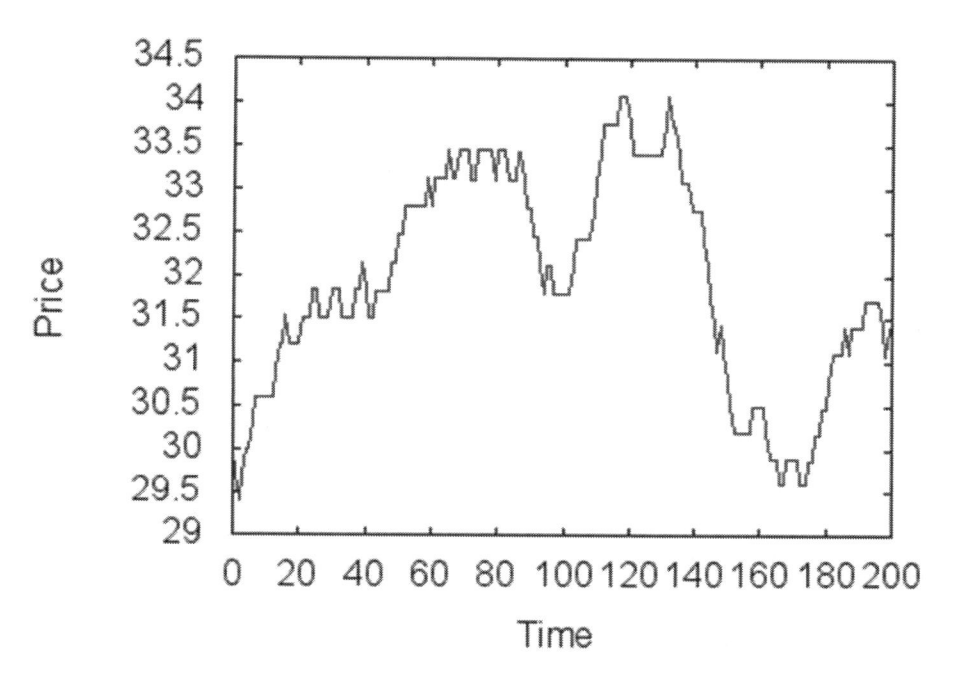

*Figure 13-5. A random walk generated by the RandomWalkModel class with starting price $30*

## Conclusion

You saw in this chapter a few examples of Monte Carlo techniques, which can be used to solve complex problems through simulation of random events. These methods are based on the use of pseudo-random values as a tool for the probabilistic analysis of events. Such models also support the simulation of complex mathematical models, including the evolution of stock prices, as well as their options and related derivative instruments.

In this chapter, you learned about the building blocks of Monte Carlos methods. First, you saw to generate pseudo-random numbers using the C++ standard library. The random numbers can also be generated according to a pre-defined probability distribution. The C++ standard library contains some of the best-known probability distributions, which makes it easy to integrate these features into user applications.

You also saw to implement a simple random walk model. In a random walk, values change by small increments in either negative or positive directions. The random walk model can be used to analyze several financial instruments, ranging from fixed income instruments to equities and derivatives.

The next chapter will cover additional library functions and classes that are commonly used to analyze and develop solutions for options and derivatives.

# CHAPTER 14

# Using C++ Libraries for Finance

Writing good financial code is a difficult task, one that cannot be done in isolation. As a software engineer, you frequently need to collaborate with others to achieve your development goals. You also need to use code that has been written by other groups. In particular, developers are constantly using libraries created by other companies or open source projects. Integrating these libraries into your own work is a major step to improve productivity.

In the world of quantitative finance, a number of C++ libraries have been used with great success. This chapter reviews some of these libraries and discusses how they can be integrated into your own applications. Some of the topics covered in this chapter include the following:

- *Boost introduction*: The boost repository provides access to many C++ libraries that are based on templates for higher efficiency. You will learn how to install and use boost, as well as integrate particular libraries in the repository to your own applications.

- *Boost odeint*: The odeint library is a well-tested and efficient set of algorithms for the solution of ordinary differential equations (ODEs). You will learn about the different algorithms contained in odeint and the different situations in which they can be employed.

- *QuantLib*: The QuantLib library has been designed as a repository for quantitative algorithms and assorted utilities for financial applications. Many parts of this code can be used to simplify the process of analyzing options and derivatives. You will learn how to use this library and see a few of the most commonly used classes and algorithms that are available in the QuantLib repository.

## Boost Libraries

In the last few years the boost project has become well known for providing high-quality libraries for C++ applications. As a result, the boost project is now the de facto repository for extensions to the STL. In fact, many of the libraries that started as part of the boost repository have been incorporated to the C++ standard, including, for example, `std::shared_ptr` and `std::unique_ptr`. A few of the developers working on boost libraries have also become part of the standard C++ committee.

The boost project focuses on using the modern features of the C++ language, including, but not exclusively, the employment of templates for high performance. Many of the libraries included in boost provide template-based interfaces that make the resulting system much more flexible. For example, different algorithms can be specialized at the template level, so that you can combine different algorithms through the use of templates, when deciding on the optimal techniques to solve a specific problem. This is a much more adaptable strategy, rather than relying on decisions made by library designers.

Note that boost is not a finance library. Instead, it provides a large number of features that are packaged in a few separate libraries. However, many of the components have direct use in the implementation of financial applications. Its components can be used to perform and simplify several tasks, such as:

- *Solving ODEs:* Ordinary differential equations appear frequently in the solution of numerical problems in the area of finance. As you have seen, to solve some option analysis models, it is necessary to efficiently compute the value of ODEs. The odeint library gives you access to such functionality, as you see in the next section.

- *uBLAS:* The Basic Linear Algebra System library provides a C++ interface to an advanced linear algebra library. uBLAS can be used to support more complex matrix-related code, as well as the solution of systems of equations.

- *Multi-array:* Many applications require the use of multi-dimensional arrays when working in areas such as 3D animation, weather predictions, etc. The multi-array library provides an easy interface for the creation and manipulation of arrays that can be indexed using multiple indices.

- *Managing file and directories:* The `<filesystem>` header file contains a set of templates that can be used to manage files and directories. It handles different operating systems, so that you don't need to rely on system-specific libraries for common file-based operations.

■ **Note** The `filesystem` library is scheduled to become part of the C++ standard library in the next few years. Meanwhile, boost can be used to gain access to this functionality.

The boost repository contains a large set of useful libraries for C++ development, including the ones listed previously. In its current version, there are 136 libraries that cover all types of tasks needed in modern programing. Table 14-1 shows a list of commonly used libraries contained in the boost project repository, including a quick explanation of their usage.

*Table 14-1. List of Commonly Used Boost Libraries*

| Library | Description |
| --- | --- |
| odeint | Implements algorithms to solve ordinary differential equations (ODEs). |
| filesystem | A set of classes to manipulate files and directories in an OS-independent way. |
| Multi-array | Provides arrays with multiple dimensions; useful for scientific code. |
| MPI | Implements the Message Passing Interface, a standard for parallel processing. |
| Math | A set of mathematical functions not included in the standard library. |
| Graph | A library that extends the STL and provides containers and algorithms to handle graphs. |
| Functional | Provides templates that simplify functional programming techniques. |
| Algorithm | A set of generic algorithms that extends the algorithm header in the STL. |
| uBLAS | A modern C++ implementation of BLAS (Basic Linear Algebra Subprograms). |
| Variant | A container that safely stores a union container, capable of storing different data types. |
| Sort | Implements several sorting strategies using templates for high performance. |
| Regex | Provides support for regular expressions in C++. |
| Python | A set of templates and classes that allows interaction between Python and C++ code. |

## Installing Boost

The first step in using the boost libraries is to install them on your machine. Being an open source repository, boost packages are made available through the web and mirrored in several web sites. The canonical web site for the repository is www.boost.org, where you can find instructions for installing boost in several architectures and operating systems.

The most common way to install boost is to download the compressed file containing the headers and source files. Once the files are uncompressed, you can use the main installation script that is provided, bootstrap.sh, to build and install the software on the desired path in the local disk.

Another way to install boost libraries is to use third-party installers or package managers. For example, if you use Linux, it is possible to install boost as a package using the local package manager, such as dpkg on Debian systems. On Windows systems, you can also install cygwin, which contains a package manager with several common C++ programming packages, including the boost libraries.

Installing from source is also easy. You just need to unzip the source files into a location and use that directory as the include path for the compilation process. An advantage of boost is that most of the libraries are implemented as header files (this is also true for most of the STL). Therefore, there is no need for any compilation. A few libraries, however, require a compilation step that can be performed using the bootstrap script. You will need the build step if you need to use one of the following libraries:

- Boost.Filesystem
- Boost.IOStreams
- Boost.ProgramOptions
- Boost.Python
- Boost.Regex
- Boost.Serialization
- Boost.Signals
- Boost.Thread
- Boost.Wave

Boost libraries are built using a C++ build system called bjam. The build script will try to find bjam in your machine or build it. You can also download bjam from its binary distribution located in boost.org/build.

In the next few sections, you will see how to use a few libraries available from boost. First, you will see how to solve ordinary differential equations with the odeint library.

## Solving ODEs with Boost

In the previous chapter, you saw how ordinary differential equations (ODEs) can be implemented directly using C++ code. Due to how options are defined and represented, ODEs models arise naturally in the design of financial algorithms. As a result, being able to quickly implement such methods is a great advantage for the quantitative software developer. Moreover, it is much easier to reuse an ODE implementation that has already been reviewed and thoroughly tested, especially considering that numerical errors are hard to catch in many cases.

One of the components of the boost repository, the odeint library, deals specifically with ODEs. With odeint, you can more easily create code to integrate ODEs, choosing from a number of different algorithmic strategies. Figure 14-1 shows a screenshot of the current web page for the odeint web site, where its repository is maintained.

# CHAPTER 14 ■ USING C++ LIBRARIES FOR FINANCE

**ODEINT** is a modern C++ library for numerically solving Ordinary Differential Equations. It is developed in a generic way using Template Metaprogramming which leads to extraordinary high flexibility at top performance. The numerical algorithms are implemented independently of the underlying arithmetics. This results in an incredible applicability of the library, especially in non-standard environments. For example, odeint supports matrix types, arbitrary precision arithmetics and even can be easily run on CUDA GPUs - check the **Highlights** to learn more.

Moreover, **ODEINT** provides a comfortable easy-to-use interface allowing for a quick and efficient implementation of numerical simulations. Visit the impressively clear 30 lines **Lorenz example**.

**ODEINT** is a header only C++ library and the full source code is available for **download.** distributed under the highly liberal **Boost Software License.** Hence, **ODEINT** is free, open source and can be used in both non-commercial and commercial applications.

**ATTENTION:** Boost is changing its structure and is being modularized from one large svn repository to many small git repository. Due to this modularaization we need to change the file structure of odeint. All headers are now located in a subdirectory include. If you do not building odeint via bjam from this repository you only need change your include path.

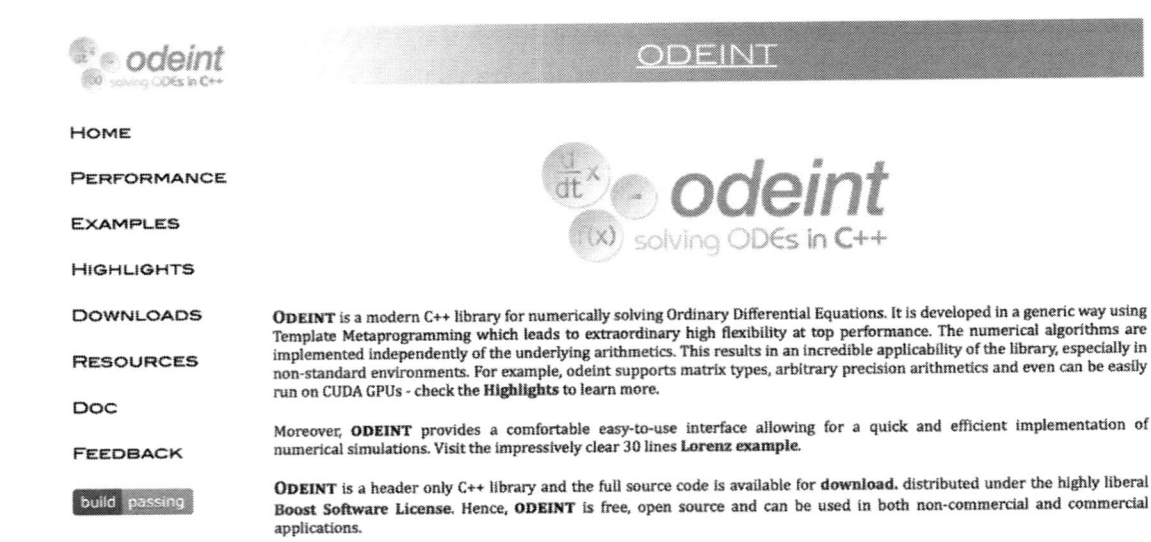

*Figure 14-1. Web site of the odeint library, where you can download its latest version*

Table 14-2 presents a quick list of the integration techniques available when using odeint. Some of these techniques have been discussed in the previous chapter. Others are variations of the best-known algorithms and can provide performance advantages for use in particular applications.

*Table 14-2. List of Integration Techniques Available When Using odeint*

| Class Name | Description |
| --- | --- |
| euler | Original Euler's algorithm to solve ODEs |
| runge_kutta4 | Uses the Runge-Kutta method, with fourth-order approximation |
| runge_kutta_cash_karp54 | Runge-Kutta method |
| runge_kutta_fehlberg78 | Variation of Runge-Kutta that uses the Fehlberg algorithm |
| adams_moulton | A multi-step algorithm for solving ODEs |
| dense_output_runge_kutta | An implementation of Runge-Kutta that uses dense output |
| bulirsch_stoer | Based on the Bulirsch-Stoer algorithm, provides higher accuracy in the solution of complex ODEs |
| implicit_euler | A variation of Euler's algorithm in which the equation is given in implicit form and requires the use of the associated Jacobian |

The algorithms made available in the odeint library are implemented as separate template classes. Each class corresponds to an algorithm or algorithmic concept. The odeint library contains a set of integration methods that can be parameterized using the provided templates. These templates make it possible to use different strategies through the combination of the given algorithms and concepts.

CHAPTER 14 ■ USING C++ LIBRARIES FOR FINANCE

One of the basic types of strategies classes available in odeint is a stepper. A *stepper* is used to navigate through the solution space of the given ODE. This is an important concept because ODEs are solved interactively, and the step size and direction determine how a particular solution strategy will behave. Depending on the type of stepping strategy used, the resulting algorithm can perform a calculation that is faster or more accurate. Here are the known stepper types provided by odeint:

- runge_kutta4
- euler
- runge_kutta_cash_karp54
- runge_kutta_dopri5
- runge_kutta_fehlberg78
- modified_midpoint
- rosenbrock4

## Solving a Simple ODE

In this section, you will see how to use the concepts described previously to solve a simple ODE in the standard form given by:

$$y' = f(x,y)$$

Here, $y$ is a function of $x$, $y'$ is the first derivative of $y$, and $f(x, y)$ is a general equation that may depend both on $x$ and $y$.

To use odeint, the first step is to include the main header file containing this library, with:

```
#include <boost/numeric/odeint.hpp>
```

To solve any ODE, you need first to determine the $f(x, y)$ part of the system, that is, the right side of the ODE equation. In this example, you will solve for the simple equation

$$y' = \frac{3}{2.5x^2} + \frac{y}{3/2x}$$

This is done in the following code fragment:

```
#include "boosttest.hpp"

#include <iostream>
#include <boost/array.hpp>

#include <boost/numeric/odeint.hpp>

//
// This is the equation at the right side of the ODE   y' = f(x,y)
// It is evaluated in the inner steps of the algorithm.
//
```

227

## CHAPTER 14 ▪ USING C++ LIBRARIES FOR FINANCE

```
void right_side_equation(double y, double &dydx, double x)
{
    dydx = 3.0/(2.5*x*x) + y/(1.5*x);
}
```

An optional feature of odeint algorithm is the use of an observer. The observer is a function that can be used to inspect each step of the algorithm. Using this information, you can record the progression of the solution, or you can perform more complex analysis if necessary. In this example, the observer simply prints the output, which will later be used to plot the convergence of the solution.

```
// this function simply prints the current value of the interactive
// solution steps.
void write_cout( const double &x , const double t )
{
    cout << t << '\t' << x << endl;
}
```

Next, you need to define the stepper algorithm. In this case, the runge_kutta_dopri5, a basic stepper based on the Runge-Kutta method, was selected. This can be done with a simple typedef to define the stepper_type.

```
// A stepper based on Runge-Kutta algorithm.
// the state_type use is 'double'
typedef runge_kutta_dopri5<double> stepper_type;
```

Finally, the main function is used to integrate the ODE under the given initial conditions. The task is performed by the integrate_adaptive function, which takes as parameters the stepper, the ODE defining equation, state and step parameters, and a function that prints the intermediate results.

```
// This solves the ODE described above with initial condition x(1) = 0.
//
int main()
{
    double x = 0.0;
    auto n = integrate_adaptive(
           make_controlled(1E-12, 1E-12, stepper_type()),  // instantiate the stepper
           right_side_equation,            // equation
           x,                              // initial state
           1.0 , 10.0 , 0.1 ,              // start x, end x, and step size
           write_cout );
    cout << " process completed after " << n << " steps \n";
    return 0;
}
```

I ran this code and used the output of the observer function to plot the convergence of the results found by the ODE solver. The plot, displayed in Figure 14-2, shows how solution values change as you move from 1.0 to 10.0 in the solution space.

CHAPTER 14 ■ USING C++ LIBRARIES FOR FINANCE

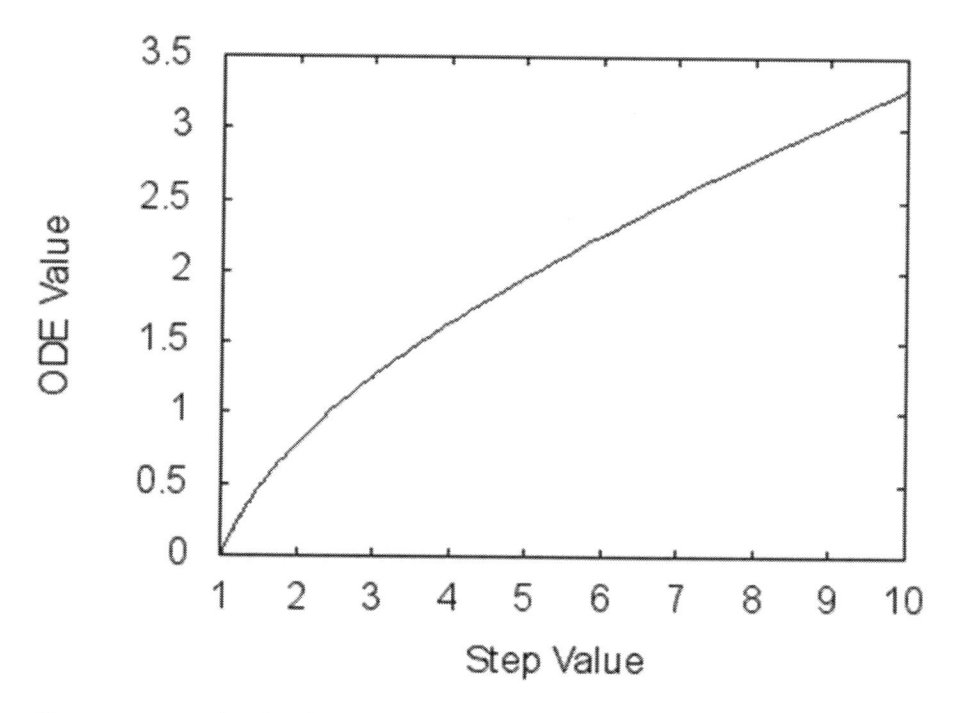

*Figure 14-2. Results of the* `integrate_adaptive` *function from the odeint library*

## The QuantLib Library

The second example of a library that is used in quantitative finance and options analysis is the QuantLib library. QuantLib is a well-established repository of quantitative code for C++. The library has been tested and used by many developers, which means that you can take advantage of the hard work that went into creating and testing the algorithms.

Being an open source project, QuantLib is free and can be used by anyone by just downloading and building the source code. The project also accepts contributed code, which means that many people can fix bugs and participate in the improvement of the library.

The QuantLib contains a wide assortment of classes that simplify certain tasks that are necessary in quantitative algorithms for finance. A few areas covered by QuantLib are the following:

- *Date handling*: Many algorithms for options and derivative analysis are based on dates. Therefore, accurate information about trading dates, holidays, and other calendar-specific events are very important for the correct results of such algorithms. QuantLib provides a number of classes that encapsulates the concepts needed for data handling in financial applications.

- *Design patterns*: The QuantLib library puts a lot of effort in following well-established design patterns. Most algorithms use design patterns that make them easier to understand and to maintain. For this reason, QuantLib has a rich implementation of common design patterns, including Singleton, Observer, Singleton, Composite, and others.

CHAPTER 14 ■ USING C++ LIBRARIES FOR FINANCE

- *Monte Carlo methods*: A few of the classes provided by QuantLib are used to simplify the implementation of Monte Carlo methods. These classes make it easier to create, for example, random paths for financial instruments, as well as similar models based on Brownian motion.
- *Pricing engines*: Another area that is covered by QuantLib is the implementation of efficient pricing engines for options and derivatives. The library provides several techniques for options pricing, which are carefully packaged into C++ classes. These pricing engines include barrier option engines, Asian option engines, basket option engines, and vanilla option engines.
- *Optimizers*: Another utility that is frequently employed in financial applications is an optimization engine. The QuantLib library contains a few classes dedicated to some common optimization strategies. Using such optimization algorithms, it is possible to quickly solve complex problems where the objective is to find the minimum or the maximum of a given function.

In the remaining of this section, you will see a few examples using classes from QuantLib. You will learn how to use some of the main classes available in the library and integrate them to your applications.

## Handling Dates

One of the most common tasks in financial algorithms is handling dates correctly. You saw in Chapter 3 that there are several ways to store and transform values stored as dates. The QuantLib library tries to simplify some of these tasks with the introduction of carefully designed date and time classes.

Managing holidays is one of the most difficult problems when using dates in financial applications. Since the number of trading days constitute part of the calculation, when computing the price of an option, it becomes very important to have precise representations of date intervals, considering which of those days are trading days.

First, lets consider how to use the Date class provided by QuantLib, along with some of the basic operations defined on that class. The basic way to construct an object of type Date is to pass the desired date in the day-month-year format. Here is an example:

```
Date date1(10, Month::April, 2010);
```

This would create a date representing the tenth day of April, 2010. Now, using a date created in this way, it is possible to perform operations such as addition or subtraction using the operators that have been overloaded by QuantLib.

```
void testDates()
{
    Date date(10, Month::April, 2010);
    cout << "original date: " << date << endl;

    date += 2 * Days;
    cout << "after 2 days: " << date << endl;

    date += 3 * Months;
    cout << "after 3 months: " << date << endl;
}
```

In this code, the operators are used to add a period of two days and three months, respectively, to the original date. The Days and Month identifiers are simple data types that concisely represent a time period and can be used to simplify the handling of intervals.

Another simple operation on dates is incrementing and decrementing. This allows you to quickly find the next or the previous day in a sequence, without needing to check if these dates occur in different months or years. The following code shows an example of how this works:

```
void nextAndPreviousDay()
{
   Date date(28, Month::February, 2010);
   cout << "original date: " << date << endl;

   date++;
   cout << "next day: " << date << endl;

   date--;
   cout << "previous day: " << date << endl;
}
```

Additional tools are provided to answer common questions related to dates. For example, member functions of the Date class are used to determine if a particular date occurs in a leap year, if it occurs at the end of the month, or if the date is a weekday. These are exemplified by the code in the following section.

## Working with Calendars

Another aspect of dates that causes a lot of confusion is handling local holidays. Each country has non-trading days that are determined by holidays, which also change according to the year in which they occur. To handle these issues, QuantLib provides a set of Calendar objects. These calendars are localized and can be used to determine if a particular date is a holiday.

The following example shows how to use the Calendar class in a typical C++ application:

```
void useCalendar()
{
   Calendar cal = UnitedStates(UnitedStates::NYSE);

   cout << " list of holidays " << endl;
   for (auto date : Calendar::holidayList(cal, Date(1, Month::Jan, 2010),
                                         Date(1, Month::Jan, 2012)))
   {
      cout << " " << date;
   }

   cout << " is Jan 1 2010 a business day?   "
       << cal.isBusinessDay(Date(1, Month::Jan, 2010)) << endl;
   cout << " is Jan 1 2010 a holiday?   "
       << cal.isHoliday(Date(1, Month::Jan, 2010))    << endl;
   cout << " is Jan 1 2010 end of month?   "
       << cal.isEndOfMonth(Date(1, Month::Jan, 2010)) << endl;
}
```

231

CHAPTER 14 ■ USING C++ LIBRARIES FOR FINANCE

The first line of the useCalendar function shows how to create a new calendar for a particular region. In this case, the calendar corresponds to the United States, and in particular to the New York Stock Exchange.

With this calendar loaded, it is possible to answer a number of questions about dates in the United States. For example, the next few lines show how to list all holidays with the holidayList function. The function receives as arguments the calendar and the desired start and end date. The result is a container with all the holidays for the given period.

The next few lines show how to use QuantLib Calendar object to answer a few common questions related to the day of the week and the month. The first call is to isBusinessDay, and it returns true if the given date occurs in a business day (usually Monday to Friday in most markets). The second member function is isHoliday, which returns true only if the given date is a holiday.

Finally, you can see the member function isEndOfMonth example. This function returns true if the given date occurs at the end of a month, which may be an important date in some kinds of financial contracts.

Another interesting feature of the Calendar class is that you can create and manage your own calendars. This is necessary when creating code for countries that are not already covered by the library, or when you're working on particular institutions or markets that use a distinct calendar.

The main functions to manage calendar holidays are addHoliday and removeHoliday. With these functions, you can create calendars that are specific to your needs. The following example code shows how to use them:

```
Calendar createNewCalendar()
{
    Calendar newCal = UnitedStates(UnitedStates::NYSE);

    // Remove winter holiday
    newCal.removeHoliday(Date(25, Month::December, 2016));

    // Add international workers' day
    newCal.addHoliday   (Date(1,  Month::May, 2016));

    cout << " list of holidays " << endl;
    for (auto date : Calendar::holidayList(newCal, Date(1, Month::Jan, 2016),
                                            Date(31, Month::Dec, 2016)))
    {
        cout << " " << date;
    }

    return newCal;
}
```

This function starts with the creation of a new calendar object based on the U.S. calendar, more specifically using the NYSE list of holidays. The function then proceeds to modify the original calendar, adding a common holiday and adding another so the number of holidays remains the same. The code also prints the list of holidays for the year 2016 to the standard output. Finally, the createNewCalendar function returns the newly created calendar as the result.

Another important feature of the Calendar class provided by QuantLib is the ability to determine the number of trading days between two dates. This is done using the businessDaysBetween member function, which returns the number of business days in a particular interval given to the function. A simple example can demonstrate how this function works:

```
int getNumberOfDays(Date d1, Date d2)
{
```

232

# CHAPTER 14 ■ USING C++ LIBRARIES FOR FINANCE

```
    Calendar usCal = UnitedStates(UnitedStates::NYSE);

    int nDays = usCal.businessDaysBetween(d1, d2);

    cout << " the interval size is " << nDays << endl;

    return nDays;
}
```

In the beginning, the getNumberOfDays function creates a calendar using the U.S. locale. The next step is to determine the number of business days between two given dates. Then, the function prints the value of this difference and returns that value as the final result.

## Computing Solutions for Black-Scholes Equations

The next example of QuantLib is directly related to the problem of pricing options. The main formula for pricing options is derived from the Black-Scholes differential equations. This makes it really important to have a library that can quickly solve Black-Scholes models, at least as an initial step for further analysis.

The QuantLib provides classes that are specifically designed to solve Black-Scholes models. Unlike other ODE and PDE packages that can be used to solve general differential equations (as seen in the previous section on boost), the QuantLib classes target efficient techniques to solve a single model in particular. This results in a very specialized algorithm that can be relied on for the efficient solution of Black-Scholes models.

To benefit from options models used by the QuantLib, you need to instantiate two classes:

- *A class representing the option and the associated payoff*: QuantLib provides a set of classes for this purpose. An example of such a class is PlainVanillaPayoff, which represents a common (vanilla) option and its associated payoff.

- *A class representing the pricing method*: This class encapsulates the algorithm that is used to compute the option price. This example is interested in the class representing the Black-Scholes algorithm, which is named the BlackScholes calculator.

These classes are exemplified in the following sample code, which includes a function that performs the computation and an associated test function.

First, you need to create a simple storage area, where the necessary information for the algorithm is stored. The BlackScholesParameters structure is used for this purpose. The structure contains the following fields:

- The spot price for the underlying instrument
- The strike price for the desired option
- The current interest rate
- The forward interest rate
- The volatility of the underlying instrument

The structure can be represented in the sample C++ code as:

```
struct BlackScholesParameters {
    double S0;
    double K;
    double rd;
    double rf;
```

233

CHAPTER 14 ■ USING C++ LIBRARIES FOR FINANCE

```
    double tau;
    double vol;
};
```

Based on this information, it is possible to describe the use of Black-Scholes pricing method using a C++ function. The function, called `callBlackScholes`, receives as a parameter a single reference to a structure of type `BlackScholesParameters`.

```
void callBlackScholes(BlackScholesParameters &bp)
{
    // create a vanilla option (standard option type)
    boost::shared_ptr<PlainVanillaPayoff>
        vanillaPut(new QuantLib::PlainVanillaPayoff(Option::Put,bp.K));

    // compute discount rates
    double cur_disc = std::exp(-bp.rd * bp.tau);   // current discount rate
    double for_disc = std::exp(-bp.rf * bp.tau);   // forward  discount rate
    double stdev    = bp.vol * std::sqrt(bp.tau);  // standard deviation

    BlackScholesCalculator putPricer(vanillaPut, bp.S0, for_disc, stdev, cur_disc);

    // Print options greeks
    cout << "value:" << putPricer.value() << endl;
    cout << "delta:" << putPricer.delta() << endl;
    cout << "gamma:" << putPricer.gamma() << endl;
    cout << "vega:"  << putPricer.vega(bp.tau) << endl;
    cout << "theta:" << putPricer.theta(bp.tau) << endl;
    cout << "delta Fwd:" << putPricer.deltaForward() << endl;
    cout << "gamma Fwd:" << putPricer.gammaForward() << endl;

}
```

This code works in the following way. The first instruction is necessary to create a new object describing the required option. This is done with the instantiation of an object of class `PlainVanilllaPayoff`, which indicates that the new option is of plain vanilla type (i.e., it is a standard option). The arguments passed are the type of option (either a call or a put) and the strike. These two parameters determine the type of option that you're handling, independent of the current characteristics of the market. The object of type `PlainVanillaPayoff` is stored in a `shared_ptr`, which automatically manages the lifetime of the object, cleaning up the pointer at the end of the scope of the local variable.

The next part of the `callBlackScholes` function initializes some of the parameters necessary to use the options pricer. The parameters include the current and forward discount rate, which are computed from the given interest rate using an exponential transformation. Another important parameter is the standard deviation, which measures the volatility of the underlying instrument.

Once the parameters for the options pricing model are available, you can instantiate the `BlackScholesCalculator` class, passing as parameters the object that describes the option, the current price, and the other parameters discussed previously.

Using the object of type `BlackScholesCalculator`, you can retrieve important information about the option price. The most important information is clearly the value of option at a particular date, returned by the member function value. The option Greeks also provide key information that can be used to make decisions about the option. The Greeks calculated by the `BlackScholesCalculator` include the following:

- *The delta*: Represents the marginal change in value with respect to the price of the underlying.

- *The gamma*: Represents the marginal change in delta with respect to the price of the underlying.
- *The vega*: Represents the marginal change in value with respect to the change in volatility.
- *The theta*: Represents the marginal change in value with respect to the change in remaining time.

You can test this code using a function that uses a few common values for each of the parameters and calls the callBlackScholes function. Here is an example of how this can be done:

```
void testBlackScholes()
{
   BlackScholesParameters bp;

   bp.S0 = 95.0;      // current price
   bp.K  = 100.0;     // strike
   bp.rd = 0.026;     // current rate of return
   bp.rf = 0.017;     // forward rate of return
   bp.tau= 0.62;      // tau (time greek)
   bp.vol= 0.193;     // volatility

   callBlackScholes(bp);
}
```

## Creating a C++ Interface

Based on the previous functions, it is easy to create a generic class that encapsulates a vanilla Black-Scholes pricing strategy. I called this class BlackScholesPricer, and it presents a simple interface that can be called without external references to QuantLib.

The class declaration contains a set of parameters that will be used in the constructor, as shown in the next code listing.

```
class BlackScholesPricer {
public:
   BlackScholesPricer(bool call, double price, double strike, double tau, double r, double fr,
      double vol);
   BlackScholesPricer(const BlackScholesPricer &p);
   ~BlackScholesPricer();
   BlackScholesPricer &operator=(const BlackScholesPricer &p);

   double value();

   double delta();
   double gamma();
   double theta();
   double vega();
private:
   double m_price;
   double m_strike;
   double m_tau;
```

## CHAPTER 14 ■ USING C++ LIBRARIES FOR FINANCE

```
    double m_rate;
    double m_frate;
    double m_vol;
    double m_isCall;

    boost::shared_ptr<QuantLib::BlackScholesCalculator> m_calc;
};
```

The constructor for BlackScholesPricer is responsible for initializing all the parameters with the passed arguments. Inside the constructor, you can see the code that initializes the payoff class. The option payoff can be a put or a call, depending on the value of the first parameter.

Later, you will see these parameters being used to create a new BlackScholesCalculator object. This object is stored in a shared pointer so that it can be used to answer questions about the model.

```
BlackScholesPricer::BlackScholesPricer(bool call, double price, double strike, double tau,
double r, double fr, double vol)
:m_price(price),
m_strike(strike),
m_tau(tau),
m_rate(r),
m_frate(fr),
m_vol(vol),
m_isCall(call)
{
    boost::shared_ptr<QuantLib::PlainVanillaPayoff>
        m_option (new QuantLib::PlainVanillaPayoff(
                    call ? QuantLib::Option::Call : QuantLib::Option::Put, strike));

    // compute discount rates
    double cur_disc = std::exp(-m_rate  * m_tau);   // current discount rate
    double for_disc = std::exp(-m_frate * m_tau);   // forward  discount rate
    double stdev    = m_vol * std::sqrt(m_tau);     // standard deviation

    m_calc.reset(new QuantLib::BlackScholesCalculator(m_option, m_price, for_disc, stdev,
    cur_disc));
}

BlackScholesPricer::BlackScholesPricer(const BlackScholesPricer &p)
:m_price(p.m_price),
m_strike(p.m_strike),
m_tau(p.m_tau),
m_rate(p.m_rate),
m_frate(p.m_frate),
m_vol(p.m_vol),
m_isCall(p.m_isCall),
m_calc(p.m_calc)
{}

BlackScholesPricer::~BlackScholesPricer() {}
```

```cpp
BlackScholesPricer &BlackScholesPricer::operator=(const BlackScholesPricer &p)
{
   if (this != &p)
   {
      m_price = p.m_price;
      m_strike = p.m_strike;
      m_tau = p.m_tau;
      m_rate = p.m_rate;
      m_frate = p.m_frate;
      m_vol = p.m_vol;
      m_isCall = p.m_isCall;
      m_calc = p.m_calc;
   }
   return *this;
}
```

Using these definitions, the following member functions can be used to provide access to the results of the pricing algorithm. They rely on the m_calc member variable, which already contains this stored information.

```cpp
double BlackScholesPricer::value() { return m_calc->value(); }

double BlackScholesPricer::delta() { return m_calc->delta(); }

double BlackScholesPricer::gamma() { return m_calc->gamma(); }

double BlackScholesPricer::theta() { return m_calc->theta(m_tau); }

double BlackScholesPricer::vega()  { return m_calc->vega(m_tau); }
```

## Complete Code

Listing 14-1 shows the BlackScholesPrices class. It shows an examples of how create an interface for the Black-Scholes component in QuantLib.

***Listing 14-1.*** Implementation File BlackScholesPrices.cpp

```cpp
#include <ql/quantlib.hpp>

#include <ql/pricingengines/blackcalculator.hpp>

//
// The BlackScholesPricer class provides an interface to the QuantLib
// pricer component
//
class BlackScholesPricer {
public:
   BlackScholesPricer(bool call, double price, double strike, double tau, double r, double fr,
      double vol);
   BlackScholesPricer(const BlackScholesPricer &p);
   ~BlackScholesPricer();
   BlackScholesPricer &operator=(const BlackScholesPricer &p);
```

## CHAPTER 14 ■ USING C++ LIBRARIES FOR FINANCE

```cpp
    double value();

    double delta();
    double gamma();
    double theta();
    double vega();
private:
    double m_price;
    double m_strike;
    double m_tau;
    double m_rate;
    double m_frate;
    double m_vol;
    double m_isCall;

    boost::shared_ptr<QuantLib::BlackScholesCalculator> m_calc;
};

BlackScholesPricer::BlackScholesPricer(bool call, double price, double strike, double tau,
double r, double fr, double vol)
:m_price(price),
m_strike(strike),
m_tau(tau),
m_rate(r),
m_frate(fr),
m_vol(vol),
m_isCall(call)
{
    boost::shared_ptr<QuantLib::PlainVanillaPayoff>
        m_option (new QuantLib::PlainVanillaPayoff(call ? QuantLib::Option::Call :
        QuantLib::Option::Put, strike));

    // compute discount rates
    double cur_disc = std::exp(-m_rate * m_tau);   // current discount rate
    double for_disc = std::exp(-m_frate * m_tau);  // forward discount rate
    double stdev    = m_vol * std::sqrt(m_tau);    // standard deviation

    m_calc.reset(new QuantLib::BlackScholesCalculator(m_option, m_price, for_disc, stdev,
    cur_disc));
}

BlackScholesPricer::BlackScholesPricer(const BlackScholesPricer &p)
:m_price(p.m_price),
m_strike(p.m_strike),
m_tau(p.m_tau),
m_rate(p.m_rate),
m_frate(p.m_frate),
m_vol(p.m_vol),
```

```
m_isCall(p.m_isCall),
m_calc(p.m_calc)
{}

BlackScholesPricer::~BlackScholesPricer() {}

BlackScholesPricer &BlackScholesPricer::operator=(const BlackScholesPricer &p)
{
   if (this != &p)
   {
      m_price = p.m_price;
      m_strike = p.m_strike;
      m_tau = p.m_tau;
      m_rate = p.m_rate;
      m_frate = p.m_frate;
      m_vol = p.m_vol;
      m_isCall = p.m_isCall;
      m_calc = p.m_calc;
   }
   return *this;
}

double BlackScholesPricer::value()
{
   return m_calc->value();
}

double BlackScholesPricer::delta()
{
   return m_calc->delta();
}

double BlackScholesPricer::gamma()
{
   return m_calc->gamma();
}

double BlackScholesPricer::theta()
{
   return m_calc->theta(m_tau);
}

double BlackScholesPricer::vega()
{
   return m_calc->vega(m_tau);
}
```

To compile this code, you need to install the QuantLib library for your platform and add that library to the project. For example, using the gcc compiler, you need to use the -lQuantLib option.

## Conclusion

Using good libraries is an important aspect of effective software development. Financial code, especially when options and derivatives are involved, requires the use of efficient and well-tested algorithms. For this reason, it is important that developers be acquainted with high-quality libraries that can be used to simplify the development process.

In this chapter, you learned about some libraries, such as boost and QuantLib, which have been successfully used to create financial applications handling options and other derivatives. The first example was from the boost repository, which contains several special-purpose libraries that use modern C++ features. The odeint library in particular, which is contained in the boost repository, can be used to simplify the computation of solutions to ODEs.

Another important library used in the financial software community is QuantLib. This open source financial library provides many useful algorithms implemented in modern, efficient C++. You saw examples of common utilities provided by QuantLib. The most common classes are for date handling. These utility classes can handle business calendars, date intervals, and sequences in a way that makes it possible to handle financial applications.

You also saw how to use QuantLib to quickly create options and derivative models. The BlackScholesCalculator class encapsulates the solution to the Black-Scholes model. This model is the basis for most techniques that can be used to analyze prices and variations of values for financial derivatives.

The next chapter will cover additional algorithms that can be used to process more complex derivatives, with special attention to credit derivatives. These algorithms with be compared and implemented in C++.

# CHAPTER 15

# Credit Derivatives

A credit derivative is a financial contract that aims at reducing credit risk—that is, the risk of default posed by contracts established with a business counterparty. These kinds of derivatives have become increasingly popular in the last decade, because they allow the hedging of complex financial positions even in industries that are not covered by mainstream markets.

As a financial software engineer, you are interested in modeling and analyzing such credit derivative contracts using programming code. Employing some of the methods developed in the previous chapters, it becomes possible to write applications that simplify the pricing and evaluation of such derivative instruments. In particular, credit derivatives can be modeled using some of the same tools that have already been discussed for the analysis of options.

In this chapter, you will learn how to create the C++ code that can be used in the quantitative analysis of credit derivative contracts. Here are some of the topics discussed:

- *General concepts of credit derivatives*: A general exposition of what credit derivatives are and the main types of derivatives commonly used in the marketplace.

- *Modeling the problem*: How to model the problems occurring in the area of credit derivatives. I'll present examples of how such derivatives can be modeled using tools that have been previously used for standard options.

- *Barrier options*: You will learn about barrier options and how they can be used to compute prices for large classes of credit derivatives. You will also see coding examples of how to handle barrier options in C++.

- *Using QuantLib for credit derivatives*: You will find a complete example of how to use the financial classes contained in QuantLib to implement derivatives-related C++ code. I will present the CDSSolver class, which implements a pricing strategy for derivatives based on barrier options.

## Introduction to Credit Derivatives

A credit derivative is a type of financial contract that protects participants from credit risk. Credit risk, in the large majority of the cases, refers to the risk of default (or lack of payment by other means) from a counterpart. For example, consider a company that creates a financial operation that is backed by an insurance contract. If this insurance contract is provided by a third party, this presents a risk of bankruptcy. The participants of this contract want to protect against the possibility of default, so companies can create a credit derivative that will pay a considerable amount of money if that the counterpart goes bankrupt. Such contracts are signed with another third party, which makes the payment if the bankruptcy occurs.

CHAPTER 15 ■ CREDIT DERIVATIVES

Credit derivatives can be classified according to several categories, which consider how the contract is structured and the participants in such a contract. Here are some of the most common types of credit derivatives that are traded in the market:

- *CDO (collateralized debt obligations)*: A CDO is a type of credit derivative where the obligations paid are collateralized based on some underlying asset. This process of collateralization creates a tiered system, where the several payers are pooled and graded according to their credit risk. Thus, financial companies can sell different tiers, ranging from the highest credit (AAA) to lower level that represent higher default risk (B+ for example).

- *CDS (credit default swap)*: A CDS allows companies to protect themselves against the default of a major market player. The buyer of a CDS makes one or more payments for a predefined period of time. If a default occurs on the covered asset, the CDS buyer is entitled to receive compensation for this credit event.

- *Credit default option:* A credit default option resembles an option contract, but the underlying corresponds to the credit default against which you are seeking protection.

- *CDN (credit linked note):* A CDN is a financial instrument that allows a particular type of credit risk to be transferred to other investors. Usually these notes are structured as bonds on lower risk assets, which are used to pay creditors if the target institution defaults.

- *CMCDS (constant maturity CDS):* A CMCDS works just like a CDS, but it has different rules for the amount of the payoff received in the case of a default. With the CMCDS, payoffs change based on considerations that are determined between the participants of the contract. For example, the payoff may be determined according to a particular interest rate index.

- *Total return swap:* This category of derivative is used to transfer financial results between two institutions according to a predefined contract. The buyer makes one or more payments, while it expects to receive the total return of a particular investment as a payoff. This allows some institutions, such as hedge funds, to receive the return of complex financial investment with the help of a second entity that transfers the financial return at the end of the covered period.

## Modeling Credit Derivatives

As you saw in the previous section, credit derivatives encompass a large number of financial products that have in common the mitigation of credit risk from one entity to another. This makes it difficult to come up with general models for such a wide class of financial instruments. In this section, you will see a few examples of C++ models applied to a few common classes of credit derivatives.

The first step in creating effective code for credit derivatives is to have a computer model for this type of security. Given the diversity of CD contracts, having a proper model becomes even more important so that other algorithms can be applied to this type of security without the need to understand the internal complexities of each type of credit derivative.

As a first step, you can define a simple class that can be used to store and manipulate the data corresponding to a credit default swap. The fields in this class represent the characteristic values that define a CDS contract. These values are the following:

- *Notional*: This represents the total value of the position encompassed by the contract. The notional is usually larger than the payments due to leverage that is allowed on derivatives contracts.

- *Spread*: The value paid by the buyer of the CDS. It may be paid in a particular schedule, or in a single payment.
- *Time period*: Defines the time period in which the CDS is valid.
- *Pay at default*: A Boolean value that determines if the payoff should be made at the time of credit default.
- *Is long*: A Boolean value that is true if the contract is being bought, and false if the contract is being sold.

In the next few sections, you will see how this information can be used to model CDS contracts with standard techniques employed in quantitative finance. In particular, I will discuss how to analyze such derivatives using the concept of barrier options. You will also see how to calculate the price for such barrier options.

# Using Barrier Options

This section I discusses how to use a technique that is frequently employed for the pricing of derivatives in general, including credit derivatives. To simplify the discussion, I use the most basic structure for a financial derivative so that you don't need to worry about complex contractual issues. However, the barrier technique described in this section can be expanded to solve a large class of commonly traded derivatives.

The first step in understanding the solution method is to define a barrier option. A *barrier option* is a special type of derivative where payoff occurs when a particular price level, or *barrier*, is crossed. This makes it different from a normal option, because common options have a payoff that depends on how much the underlying is above or below some threshold. With a barrier option, however, the payoff is paid only as the barrier is crossed.

Barrier options work well as a simple model for credit derivatives, because the credit event is frequently defined as a particular barrier. For example, if the credit event is the bankruptcy of a company, the barrier to be crossed is given by the difference between assets and liabilities in the corporation. When that barrier is breached, the company becomes insolvent and the payoff needs to be made.

There are two main types of barrier options, depending on how the barrier is considered as part of the contract:

- *Knock-in*: This is a barrier option where the payoff is given only when the barrier is touched before expiration.
- *Knock-out*: This is a barrier option where the payoff is given only when the barrier is *not* touched before expiration.

Thus, for example, a barrier option that pays when a company claims bankruptcy is a knock-in option, because the payment happens when the default barrier is reached. You can also classify barrier options according to the current value in relation to the barrier:

- *Down-option*: This is a barrier option where the barrier is below the current value of the underlying asset.
- *Up-option*: This is a barrier option where the barrier is above the current value of the underlying asset.

These two classifications can also be combined, so that you can have down-in options or up-out options. Finally, these options can be calls or puts, depending on whether you are buying the right to sell (put) or the right to buy (call) the underlying instrument.

243

CHAPTER 15 ■ CREDIT DERIVATIVES

# A Solver Class for Barrier Options

To solve the problem, a new class called CDSSolver is defined in this section. This class contains all the elements necessary to define a barrier option, along with the code that solves the pricing problem using functions and classes from the QuantLib repository. The definition of the class contains the member variables needed by the pricing algorithm:

```
class CDSSolver : boost::noncopyable {
public:

    // constructor
    CDSSolver(double val, double sigma, double divYield,
            double forwardIR, double strike, double barrier, double rebate);

    // solve the model
    std::pair<QuantLib::BarrierOption, QuantLib::BlackScholesMertonProcess>
    solve(QuantLib::Date maturity_date);

    // generate a grid
    void generateGrid(QuantLib::BarrierOption &option,
                    QuantLib::BlackScholesMertonProcess &process,
                    const std::vector<QuantLib::Size> &grid);

private:

    double currentValue;
    double sigma;
    double divYield;
    double forwardIR;
    double strike;
    double barrier;
    double rebate;
};
```

The first thing to consider when reviewing this class is that the QuantLib code also uses boost libraries for basic functionality, such as smart pointers. In this case, the CDSSolver uses boost::noncopyable as a base class, which indicates that the class cannot be copied. Therefore, no copy constructor or assignment operators are declared in CDSSolver.

■ **Note** Observe that the CDSSolver class uses shared pointers declared in boost. This is necessary because QuantLib has boost as a direct dependency, and many of the internal smart pointers are declared in this way. Remember, however, that C++11 also has its own version of shared_ptr, which is part of the standard namespace. It is important to avoid confusion between shared_ptr declared in boost and in the standard library.

There are two main member functions in the CDSSolver class. The solve function is responsible for performing the main tasks associated with the pricing of barrier options. The generateGrid evaluates the value of the barrier option at particular time points, as defined by the vector of times points passed as a parameter.

The member variables used by the `CDSSolver` class are the following:

- *currentValue*: Represents the current value of the underlying instrument.
- *sigma*: Represents the variance of the financial instrument.
- *divYield*: The dividend yield paid annually by the underlying.
- *forwardIR*: The forward interest rate, which is used to determine the return of cash that is not invested in the barrier option.
- *strike*: The strike of the barrier option, that is, the price that determines the payoff value.
- *barrier*: The price barrier that needs to be crossed to trigger the payout of the option contract.
- *rebate*: Contractual rebate defined when the barrier option is created.

These variables are later used to solve the pricing problem, as you can see in the following description of the associated code. But first, I will provide a short introduction to the classes included in QuantLib that are used solve this kind of pricing problem.

## Barrier Option Classes in QuantLib

QuantLib offers support for pricing credit derivatives and related instruments. In particular, the library contains a set of classes that can be used to price barrier options as defined in the previous section. First, I will review some of these classes, which will later be used in a complete example of how to compute prices for barrier options.

The first class of importance is the `Quote` class. A quote is defined as one or more values that determine the current price of an instrument. The `Quote` class is just the base for several classes that represent quotes for different financial instruments. In this example, I will use a `SimpleQuote` to initialize the quote for the barrier option. The following code shows how this is done:

```
Handle<Quote> quote(boost::shared_ptr<Quote>(new SimpleQuote(currentValue)));
```

This line of code uses a second class that is frequently used in QuantLib: the `Handle` class. A handle is a simple container that allows objects to be referenced and changed when necessary.

The next class used in the implementation of barrier options is `YieldTermStructure`. This class allows you to specify the yield curve currently used by the markets. The yield curve is a representation of the effective interest rates in a particular market, such as, for example, United States Treasury bonds. The curve is formed as you consider the different interest rates for each maturity period, usually measured in years. Figure 15-1 shows an example of the yield term structure for Treasury bonds.

Using the `YieldTermStructure` class in QuantLib, it is possible to store and use this information to compute barrier options. Depending on how the financial instrument is defined, such a yield term structure may be represented by several interest rates, one for each desired time horizon. The `YieldTermStructure` class is abstract and should be instantiated using one of their subclasses, which include:

- `FlatForward`: The simplest cases in which the curve is flat and no variation in interest rate is forecasted.
- `ForwardCurve`: A type of yield curve that can use different rates for each time period. This class can be used for the most common case where the interest rates for different time periods are known.

CHAPTER 15 ■ CREDIT DERIVATIVES

- `PiecewiseYieldCurve`: A yield curve in which the different segments of the curve are linearized.
- `FittedBondDiscountCurve`: A yield curve where interest rates are given indirectly, and the yield can be fitted to represent a set of bonds.

In the example code of the next section, I use the `FlatForward` class as a way to represent a simple short-term yield structure, with no variation in interest rates. More complex yield term structures can be easily accommodated by using one of the previous classes.

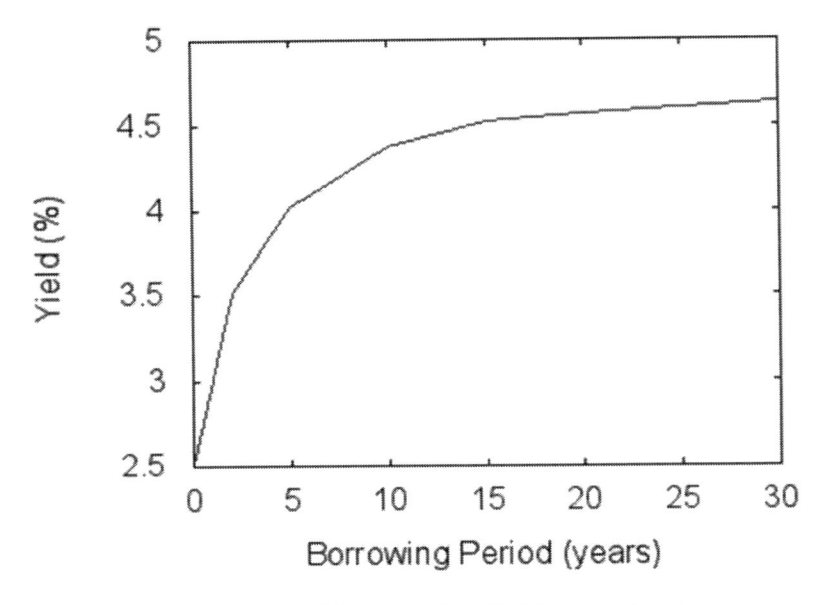

*Figure 15-1.* Example of yield structure for a U.S. Treasury bond

A similar class provided by QuantLib is the `BlackVolTermStructure`. This class represents the volatility term structure, and allows you to determine a particular curve that represents the implied volatility (also known as Black volatility, which is used in the Black-Scholes equation) for the underlying instrument. Similarly to the yield term structure, there are several options for the type of volatility term structure. They differ in the shape of the curve, as well as in the functions that can be used to represent each part of the curve. QuantLib also provides a number of classes that can be used to represent the different types of volatility term structure. Here are some of them:

- `BlackConstantVol`: Used to represent a volatility type that is constant over the whole period.
- `BlackVarianceCurve`: A type of volatility curve where different values of variance are used to determine the volatility.
- `ImpliedVolTermStructure`: A volatility term structure that is defined by the implied volatility associated with a particular instrument.
- `BlackVarianceSurface`: Defines a volatility curve based on a set of data points that define a variance surface. These values are interpolated to generate the desired variance surface.

Using the information stored in these classes, it is possible to describe the Black-Scholes model using the class `BlackScholesMertonProcess`, which is also part of QuantLib. This class receives as parameters the quote, a risk-free yield term structure, and a yield term representing the asset dividend. The class constructor also receives a volatility term structure as a parameter that describes the process.

The class `StrikedTypePayoff` is used to build complex payoffs. It also has a few useful derived classes, including the following:

- `PlainVanillaPayoff`: Represents the most common type of payoff, described by a single value and a strike.

- `PercentageStrikePayoff`: A type of payoff where the strike is given as a percentage of the underlying price, instead of as a fixed value.

- `AssetOrNothingPayoff`: A payoff that is structured as a binary decision. The results are either an asset or nothing.

- `CashOrNothingPayoff`: A payoff that is structured as a binary decision. The results are either cash or nothing.

The example code in the next section uses the `PlainVanillaPayoff` class. The constructor to this class uses as parameters the option type (put or call) and a strike.

`BarrierOption` is the central class used by QuantLib to model barrier options. This class can be used to calculate the value of a particular barrier option, given a set of parameters that represent that option.

The first parameter to the constructor of `BarrierOption` is the type of barrier option. As previously described, barrier options can be of four types—UpIn, UpOut, DownIn, and DownOut—depending on the underlying price and the type of barrier used. The next parameters are values that correspond to the barrier, the rebate, the payoff, and the exercise.

Finally, this example also uses a barrier options engine called `FdBlackScholesBarrierEngine`. This class is used as an implementation for the pricing strategy.

## An Example Using QuantLib

Using the classes presented in the previous section, it is possible to explain the implementation of the class `CDSSolver`. First, consider the first member function, called `CDSSolver::solve`. This function receives as a parameter a `Date` object that represents the maturity date of the desired barrier option.

The first step is to create a quote for the option, instantiating the `SimpleQuote` class and using the current value of the underling as its single argument. Today's date is also computed with the help of the `Date::todaysDate` member function.

Next, the code tries to instantiate the two term structure objects, one for the dividend yield and another for free cash interest rates. A volatility term structure object is also instantiated using the given volatility, which is estimated using the parameter `sigma`.

```
// solve the valuation problem using the barrier technique, from today to the maturity date
pair<BarrierOption, BlackScholesMertonProcess>
CDSSolver::solve(Date maturity_date)
{
   Handle<Quote> quote(boost::shared_ptr<Quote>(new SimpleQuote(currentValue)));
   Date today = Date::todaysDate();

   shared_ptr<YieldTermStructure>    ts1(new FlatForward(today, divYield, Thirty360()));
   shared_ptr<YieldTermStructure>    ts2(new FlatForward(today, forwardIR, Thirty360()));
   shared_ptr<BlackVolTermStructure> vs(new BlackConstantVol(today, NullCalendar(),sigma,
   Thirty360()));
```

CHAPTER 15 ■ CREDIT DERIVATIVES

The next part of the solve function is responsible for instantiating a process object, which uses QuantLib::BlackScholesMertonProcess. Such a process requires a quote object, yield term structures for interest rates and cash, and a volatility term structure that was previously created.

The function also creates two new objects: a payoff object of type PlainVanillaPayoff that represents the desired call option and a given strike. The exercise is established as a EuropeanExercise type, at the given maturity date.

```
auto process = BlackScholesMertonProcess(quote,
        Handle<YieldTermStructure>(ts1),
        Handle<YieldTermStructure>(ts2),
        Handle<BlackVolTermStructure>(vs));

  shared_ptr<StrikedTypePayoff> payoff(new PlainVanillaPayoff(Option::Type::Call, strike));
  shared_ptr<Exercise> exercise(new EuropeanExercise(maturity_date));
```

Finally, you're ready to create a barrier option object, which is an instance of QuantLib::BarrierOption. It takes as parameters the type of barrier, the barrier value, a rebate (if it is available), and the two objects previously created: payoff and exercise.

The next two steps are to create a generalized Black-Scholes object using the existing process and to set the price engine of the barrier option. The price engine algorithm is responsible for price calculation, and this example uses AnalyticBarrierEngine, which is a common algorithm available from QuantLib. The member function CDSSolver::solve will finally return a pair that contains the option and process objects.

```
auto option = BarrierOption(Barrier::Type::UpIn,
                            barrier, rebate,
                            payoff,
                            exercise);

  auto pproc = shared_ptr<GeneralizedBlackScholesProcess>(&process);

  option.setPricingEngine(shared_ptr<PricingEngine>(new AnalyticBarrierEngine(pproc)));

  return std::make_pair(option, process);
}
```

The next member function implemented in the CDSSolver class is generateGrid. This function is conceptually simple, and it just prints a grid of prices calculated from the given barrier option, using the given BlackScholesMertonProcess and a set of points that determines the option price at a particular date.

Essentially, the function assumes that the grid points are sorted and selects the maximum value. Then, for each element of the grid, a new barrier engine is instantiated and used with the existing barrier option. The price is computed using the resulting combination of option and pricing engines. The code then prints the ratio of increase for that particular point. A backward computation is also performed for comparison purposes.

```
void CDSSolver::generateGrid(BarrierOption &option, BlackScholesMertonProcess &process,
const vector<Size> &grid)
{
   double value = option.NPV();
   Size maxG = grid[grid.size()-1];   // find maximum grid value

   for (auto g : grid)
   {
```

CHAPTER 15 ■ CREDIT DERIVATIVES

```
        FdBlackScholesBarrierEngine be(shared_ptr<GeneralizedBlackScholesProcess>(&process),
        maxG, g);
        option.setPricingEngine(shared_ptr<PricingEngine>(&be));

        cout << std::abs(option.NPV()/value -1);

        FdBlackScholesBarrierEngine be1(shared_ptr<GeneralizedBlackScholesProcess>(&process),
        g, maxG);
        option.setPricingEngine(shared_ptr<PricingEngine>(&be1));

        cout << std::abs(option.NPV()/value -1);
    }
}
```

## Complete Code

This section contains the complete listing of the CDSSolver class. It contains a header file (Listing 15-1), which defines the interface for the class, and an implementation file (Listing 15-2), where the methods solve and generateGrid are implemented.

***Listing 15-1.*** Header File for the CDSSolver Class

```
//
//  CDS.hpp
//  CppOptions

#ifndef CDS_hpp
#define CDS_hpp

#include <stdio.h>

#include <utility>

#include <ql/instruments/barrieroption.hpp>
#include <ql/processes/blackscholesprocess.hpp>

//
// CDSSolver class, incorporates the solution to Credit Default
class CDSSolver : boost::noncopyable {
public:

    // constructor
    CDSSolver(double val, double sigma, double divYield,
              double forwardIR, double strike, double barrier, double rebate);

    // solve the model
    std::pair<QuantLib::BarrierOption, QuantLib::BlackScholesMertonProcess>
    solve(QuantLib::Date maturity_date);

    // generate a grid
    void generateGrid(QuantLib::BarrierOption &option,
```

```
                    QuantLib::BlackScholesMertonProcess &process,
                    const std::vector<QuantLib::Size> &grid);

private:

    double currentValue;
    double sigma;
    double divYield;
    double forwardIR;
    double strike;
    double barrier;
    double rebate;
};

#endif /* CDS_hpp */
```

Listing 15-2 shows the implementation file for class CDSSolver. It also contains a simple test stub called test_CDSSolver, which creates a new instance of CDSSolver using a few test parameters.

***Listing 15-2.*** Implementation File for Class CDSSolver

```
//
//  CDS.cpp

#include "CDS.h"

#include <iostream>

// include classes from QuantLib
#include <ql/instruments/creditdefaultswap.hpp>
#include <ql/instruments/barrieroption.hpp>
#include <ql/quotes/SimpleQuote.hpp>
#include <ql/time/daycounters/thirty360.hpp>
#include <ql/exercise.hpp>
#include <ql/termstructures/yield/flatforward.hpp>
#include <ql/termstructures/volatility/equityfx/blackconstantvol.hpp>
#include <ql/processes/blackscholesprocess.hpp>
#include <ql/pricingengines/barrier/analyticbarrierengine.hpp>
#include <ql/pricingengines/barrier/fdblackscholesbarrierengine.hpp>

using namespace QuantLib;

using std::cout;
using std::vector;
using std::pair;

using boost::shared_ptr;

CDSSolver::CDSSolver(double val, double sigma, double divYield, double forwardIR,
                     double strike, double barrier, double rebate)
:
    currentValue(val),
```

## CHAPTER 15 ■ CREDIT DERIVATIVES

```
        sigma(sigma),
        divYield(divYield),
        forwardIR(forwardIR),
        strike(strike),
        barrier(barrier),
        rebate(rebate)
{
}

// solve the valuation problem using the barrier technique, from today to the maturity date
pair<BarrierOption, BlackScholesMertonProcess>
CDSSolver::solve(Date maturity_date)
{
    Handle<Quote> quote(boost::shared_ptr<Quote>(new SimpleQuote(currentValue)));
    Date today = Date::todaysDate();

    shared_ptr<YieldTermStructure>   ts1(new FlatForward(today, divYield, Thirty360()));
    shared_ptr<YieldTermStructure>   ts2(new FlatForward(today, forwardIR, Thirty360()));
    shared_ptr<BlackVolTermStructure> vs(new BlackConstantVol(today, NullCalendar(),sigma,
        Thirty360()));

    auto process = BlackScholesMertonProcess(quote,
            Handle<YieldTermStructure>(ts1),
            Handle<YieldTermStructure>(ts2),
            Handle<BlackVolTermStructure>(vs));

    shared_ptr<StrikedTypePayoff> payoff(new PlainVanillaPayoff(Option::Type::Call, strike));
    shared_ptr<Exercise> exercise(new EuropeanExercise(maturity_date));

    auto option = BarrierOption(Barrier::Type::UpIn,
                                barrier, rebate,
                                payoff,
                                exercise);

    auto pproc = shared_ptr<GeneralizedBlackScholesProcess>(&process);

    option.setPricingEngine(shared_ptr<PricingEngine>(new AnalyticBarrierEngine(pproc)));

    return std::make_pair(option, process);
}
void CDSSolver::generateGrid(BarrierOption &option, BlackScholesMertonProcess &process,
const vector<Size> &grid)
{
    double value = option.NPV();
    Size maxG = grid[grid.size()-1];    // find maximum grid value

    for (auto g : grid)
    {
        FdBlackScholesBarrierEngine be(shared_ptr<GeneralizedBlackScholesProcess>(&process),
            maxG, g);
```

251

## CHAPTER 15 ■ CREDIT DERIVATIVES

```cpp
        option.setPricingEngine(shared_ptr<PricingEngine>(&be));

        cout << std::abs(option.NPV()/value -1);

        FdBlackScholesBarrierEngine be1(shared_ptr<GeneralizedBlackScholesProcess>(&process),
         g, maxG);
        option.setPricingEngine(shared_ptr<PricingEngine>(&be1));

        cout << std::abs(option.NPV()/value -1);
    }
}
void test_CDSSolver()
{
    // use a few test values

    double currentValue = 50.0;
    double sigma = 0.2;
    double divYield = 0.01;
    double forwardIR = 0.05;
    double strike = 104.0;
    double barrier = 85.0;
    double rebate = 0.0;

    CDSSolver solver(currentValue, sigma, divYield, forwardIR, strike, barrier, rebate);

    Date date(10, Month::August, 2016);

    auto result = solver.solve(date);

    std::vector<Size> grid = { 5, 10, 25, 50, 100, 1000, 2000 };
    solver.generateGrid(result.first, result.second, grid);
}

int main()
{
    test_CDSSolver();
    return 0;
}
```

## Conclusion

Credit derivatives are one of the most common types of derivatives traded in world markets. In this chapter, you learned a little more about such types of derivatives and how they can be modeled using C++.

I initially discussed the concept of credit derivatives and the different types of financial instruments that take part in this category of derivatives. You saw that such derivatives can be used to mitigate credit risks, such as the bankruptcy of a counterparty or the default of a loan, for example.

You also learned about techniques to model such derivatives. In particular, you saw barrier options as a simplified model that can be used to analyze the behavior of such financial instruments.

This chapter presented a complete example of credit derivatives through the use of barrier options using QuantLib classes. The QuantLib repository contains a number of algorithms that are readily available to analyze credit derivatives. In particular, these classes can be used to determine the fair price of certain types of derivatives.

Another task that is frequently necessary when dealing with options and derivatives is the processing of input and output data in common formats. The most popular format for this type of application is based on XML. However, some other formats offer advantages as well. In the next chapter, you will learn about different strategies to process financial data and the common formats used to transfer such information across applications.

# Index

## A

add function, 144
addObserver function, 94, 96
Algorithms, STL
    concepts, 115
    copying container data, 121
    find_if function, 124
    finding elements, 123
    find_value function, 123
    option data, 124
        is_expiring, 125
        StandardOption, 126
        syntax, 125
    sort function
        compute frequency, 119
        date class, 118
        date object, 118
        frequency_test, 121
        functional operator, 117
        list, 116
        vector count, 120
    types, 116

## B

Basic Linear Algebra Subprograms (BLAS), 154
Bernoulli distribution, 214
Bind function
    SimpleOption class, 134
    std::bind function, 133
    std::plus function, 134
Binomial model, 190
    algorithm, 191
    American-style option, 195
    binomial tree, 191
    call option price, 190
    expressions, 190
    implementation, 192
Black-Scholes-Merton equations, 23–24
BlackScholesMethod class, 22
    header file, 202
    implementation file, 202–205
Black-Scholes model, 175, 189
    BlackScholesMethod class, 198
    forward method, 198
    parameters, 197
    PDE, 197
buildAdjacencyMatrix function, 56

## C

Call_greater_than_100 function, 124
CDSSolver class
    header file, 249–250
    implementation file, 250–252
C++ libraries
    boost libraries
        boosttest.hpp, 227
        files and directories, 224
        installation, 225
        integrate_adaptive function, 229
        integration techniques, 226
        multi-array, 224
        solving ODEs, 224–225
        std::shared_ptr, 223
        stepper, 227
        stepper_type, 228
        types, 224
        uBLAS, 224
        web site screenshot, 225
    QuantLib
        addHoliday and removeHoliday, 232
        benefits, 233
        BlackScholesCalculator, 234
        Black-Scholes models, 233
        BlackScholesParameters structure, 233
        BlackScholesPricer class, 237
        BlackScholesPricer interface, 235
        Calendar class, 231
        Calendar objects, 231
        callBlackScholes function, 234–235
        date handling, 229–230

# INDEX

C++ libraries (cont.)
  design patterns, 229
  getNumberOfDays function, 233
  incrementing and decrementing, 231
  Monte Carlo methods, 230
  optimizers, 230
  pricing engines, 230
  useCalendar function, 232
Collateralized debt obligations (CDO), 242
Constant maturity CDS (CMCDS), 242
Credit default swap (CDS), 242
Credit derivative
 barrier option, 241
  CDSSolver class, 244
  down-option, 243
  knock-in, 243
  knock-out, 243
  up-option, 243
 CDN, 242
 CDO, 242
 CDS, 242
 CMCDS, 242
 credit default option, 242
 general concepts, 241
 problem modeling, 241
 QuantLlib, 241
  barrier option, 248
  BlackScholesMertonProcess, 247–248
  BlackVolTermStructure class, 246–247
  CDSSolver class, 247, 248
  Handle class, 245
  PlainVanillaPayoff class, 247
  Quote class, 245
  SimpleQuote class, 247
  StrikedTypePayoff class, 247
  YieldTermStructure class, 245–246
 total return swap, 242
Credit linked note (CDN), 242

# D

Date and time handling
 addTradingDays, 38
  file implementation, 40–48
  interface, 39–40
  isTradingDay function, 38
  m_day, 37
  m_holidays, 37
  monthsWithThirtyOneDays, 38
  m_weekDay, 37
  operations implementation, 36
  operator++ function, 37
 DateCompact class
  bool DateCompact, 49
  implementation, 50
  int DateCompact, 49
  int day() method, 48
  interface, 49
  int month() method, 48
  int year() method, 48
  testing, 53
  void DateCompact, 49
  void setDay(int d) function, 48
  void setMonth(int m) function, 48
  void setYear(int y) function, 48
 date operations, 36
DEFAULT_NUM_INTERVALS constant, 170
DEMathFunction, 178
Derivatives
 arbitrage, 21
 binomial trees
  backward phase, 25
  forward phase, 25
  payoff phase, 25
 Black-Scholes-Merton equations, 23–24
 code implementation, 32
 collateralized debt obligation, 21, 22
 computeRandomStep function, 28
 credit default swap, 21
 energy derivative, 21
 FX derivatives, 21, 23
 inflation derivative, 21
 interest rate derivatives, 21
 Monte Carlo models, 25
 numerical methods, 24
 profit chart, 33
 RandomWalkGenerator class, 27, 29
 random walk model, 20
 standard template library
  algorithms, 26–27
  containers, 26
  iterators, 26
Dictionary class
 adjacency matrix, 55
 buildAdjacencyMatrix function, 56
 diffByOne function, 56
 elemPosition function, 55
 implementation, 59
 interface, 59
 m_adjacentList variable, 54
 m_valuePositions, 54, 55
 m_values vector, 54, 55
 m_wordSize variable, 54
 push_back, 55
 recoverPath method, 58
 shortest_path function, 58
 StringProduction class, 57
  implementation, 62
  interface, 61
  main Function, 65

■ INDEX

test code, 66
diffByOne function, 56
Differential equations
    basic techniques, 175
    classification, 176
    definition, 175
    Euler's method, 175
    ODEs, 175–176
    order, 176
    PDEs, 175
    Runge-Kutta method, 175

■ E

elemPosition function, 55
Euler's method, 175
    DEMathFunction, 178–179
    EulerMethodSampleFunction class, 180
    EulersMethod class, 179
    incremental point, 180
    initial coordinate values, 180
    ODEs, 177
    parameters, 179
    sequential steps, 177
    test code, 181
Exchange Traded Funds (ETFs), 5

■ F

Factory design pattern
    DataSource class, 88
        declaration, 88
        implementation, 89–90
    object creation, 87
    private modifier, 88
Functional programming, 68
    advantages, 128
    bind function, 133
    Functional.cpp files, 138
    Functional.hpp files, 137
    function objects
        comparison object, 131
        operator(), 130
        OptionComparion class, 129
    lambda function, 135
    plus function, 132
    std::vector, 132
    templates list, 131
FX derivatives, 19

■ G

generateWalk() function, 27
Generic/template-based programming, 68

getFunctionRoot function, 165, 167
get_normal_observations function, 216
get_uniform_int function, 212

■ H

Horner's method, 163

■ I, J, K

int day() method, 48
int month() method, 48
int year() method, 48
isTradingDay function, 38, 74

■ L

Lambda capture by reference, 136
Lambda capture by value, 136
Lambda function
    lambda capture, 136
    syntax, 135
    test_use_function, 137
    use_function, 137
Linear algebra algorithms
    BLAS, 154
    LA functions, 143
    LAPACK, 154
    matrix implementation, 143
        add member function, 152
        definition, 148
        features, 148
        matrix class declaration, 149
        m_rows vector, 150–151
        multiply member function, 153
        numRows member function, 154
        operator[] member function, 151
        product operation, 153
        rectangular matrix, 150
        subtract operation, 152
        trace operation, 151
        transposition, 151
    uBLAS libraries
        matrix object, 155
        preMultiply function, 155
        prod function, 156
        vector, 155
    vector operations, 143
        Header File LAVectors.hpp, 156
        Implementation File LAVectors.cpp, 157
        mathematical operations, 144
        scalar operations, 144
        vector-to-vector operations, 146
Linear algebra package (LAPACK), 154
linear_congruential_engine, 209

257

■ INDEX

## ■ M

Monte Carlo methods
    advantages, 207
    bonds analysis, 208
    fixed income investments, 208
    market analysis, 207
    options pricing, 208
    Portfolio analysis, 208
    probability distributions, 207
    random number generation, 207
        algorithms, 209
        engine instantiation, 209
        generator instantiations, 210
        linear_congruential_engine, 209
        Mersenne twister, 208
        probabilitydistribution (*see* Probability distribution)
        subtract with carry, 209
    random walk, 207, 218
    stochastic models, 207
    trade strategy analysis, 208
Monte Carlo model, 19, 25
multiply function, 145, 153

## ■ N

Networks
    dictionaryclass (*see* Dictionary class)
    string-production, 54
    word production, 53
num_customers_experiment function, 214–215
Numerical programming algorithms
    integration algorithms, 161
    mathematical function representation, 161
        Horner's method, 163
        MathFunction, 162
        PolynomialFunction, 162–163
    numerical examples in C++, 161
    rootfinding (*see* Root finding algorithms)

## ■ O

Object-oriented programming
    C++ concepts, 67
    class hierarchies, 80–81
    design patterns
        factory method, 86
        observer, 86
        singleton, 86
        visitor, 86
    encapsulation, 69
        CDS contract, 70–71
        features, 69
    inheritance, 69, 72, 81
    object composition, 82–83
    polymorphism, 69
        abstract functions, 78–80
        CDSContract class, 73
        isTradingDay member function, 74
        useContract function, 74
        virtual function, 74
        virtual keyword, 72–73, 76
        virtual destructor, 77–78
        virtual table mechanism, 75
    problem partitioning, 67
    problem solving, 67
    reusing components, 67
Object-oriented programming, 68
Observer design pattern
    addObsever function, 94
    header file, 96
    implementation file, 97
    Observer class, 94
    removeObserver function, 94
    simplified scheme, 94
    trade observer, 95
    TradingLedger class, 95
    triggerNotification function, 94, 96
Operator++ function, 37
Option Greeks, 6
Options pricing
    binomial model
        algorithm, 191
        American-style option, 189, 195
        binomial tree, 191
        call option price, 190
        expressions, 190
        implementation, 192
    binomial trees, 189
    Black-Scholes model
        BlackScholesMethod class, 189, 198
        forward method, 198
        parameters, 197
        PDE, 197
    implementation strategies, 189
    lattice models, 189
Options processing
    American options, 4
    break-even price, 6
    C++ classes
        assignment operator, 11
        copy constructor, 11
        CppClass, 11
        destructor, 11
        GenericOption class, 12–13
        profitAtExpiration, 12–13
        profit chart, 17
        valueAtExpiration, 12
    C++ programming language

availability, 8
expressiveness, 10
performance, 9
standardization, 9–10
definition, 1–2
delta hedging, 1
design patterns
factory method, 85–86
observer, 85–86
OO programming techniques, 86
overview, 85
singleton, 85–86
visitor, 86
European options, 4
expiration, 3
features, 2
fundamental strategies, 1
Greeks, 1, 6
intrinsic value, 5
profit chart, 3
sellers, 4
settlement, 4
strike price, 3
trading, 2
at the money (ATM), 5
commodities, 5
common stock, 4
currencies, 4
ETFs, 5
futures, 5
indices, 4
in the money (ITM), 5
out of the money (OTM), 5
Ordinary differential equations (ODEs), 225
analytical methods, 176
Euler's method, 178
numerical methods, 176
Runge-Kutta method, 182

# P, Q

Partial differential equations (PDEs), 175
Placeholder arguments, 134
Poisson distribution, 214
preMultiply function, 155
Probability distribution
coin_toss_experiment function, 213
exponential distribution, 211
get_normal_observations function, 216
get_uniform_int function, 212
histogram values, 217
normal distribution, 211
num_customers_experiment function, 214–215

num_experiments parameter, 214
std:uniform_int_distribution template, 212
processNewTrade function, 95
prod function, 156

# R

RandomWorkModel class
class interface, 219
getWalk function, 218, 220
member variables, 219
m_numSteps variable, 220
RandomWalkModel, 219–220
test_random_walk function, 220–221
recoverPath method, 58
removeObserver function, 96
Resource Acquisition Is Initialization (RAII), 9
Root finding algorithms, 161
Newton's method
derivative class, 168
derivative objects, 168
getFunctionRoot function, 165, 167
NewtonMethod class, 165
SampleFunction class, 167, 168
sequence values, 169
SimpsonsIntegration class, 170–173
Runge-Kutta method, 175, 228
class, 183, 185–187
DEMathFunction, 183
fourth order approximation, 184
ODEs, 182
RungeKuttaSampleFunc, 185
Taylor series, 182
test function, 185

# S

shortest_path function, 58
Singleton design pattern
clearing house, 90–91
Company CEO, 91
memory allocator, 91
operating system, 91
root window, 91
Standard template library (STL), 10
algorithms, 26–27 (*see also* Algorithms, STL)
containers, 26, 109–110
iterators, 26
Stepper types, 227
Structured programming, 68
swap member function, 151

259

# INDEX

## T

Taylor method, 182
Templates
    compile-time polymorphism, 102–103
    container objects, 102
    data containers, 109
    definition, 101
    generic functions, 104–105
    header file, 112
    instantiation, 111
    recursive function, 102
        implementation, 106
        integer values, 107
        syntax, 106
        template classes, 108
    smart pointers, 102, 110–111
test_normal function, 216
test_random_walk function, 220–221
triggerNotification function, 94, 96
Trinomial model, 190

## U

useContract function, 74

## V, W, X, Y, Z

virtual function, 74
void setDay(int d) function, 48
void setMonth(int m) function, 48
void setYear(int y) function, 48

# Get the eBook for only $5!

Why limit yourself?

Now you can take the weightless companion with you wherever you go and access your content on your PC, phone, tablet, or reader.

Since you've purchased this print book, we're happy to offer you the eBook in all 3 formats for just $5.

Convenient and fully searchable, the PDF version enables you to easily find and copy code—or perform examples by quickly toggling between instructions and applications. The MOBI format is ideal for your Kindle, while the ePUB can be utilized on a variety of mobile devices.

To learn more, go to www.apress.com/companion or contact support@apress.com.

## Apress®
### THE EXPERT'S VOICE™

All Apress eBooks are subject to copyright. All rights are reserved by the Publisher, whether the whole or part of the material is concerned, specifically the rights of translation, reprinting, reuse of illustrations, recitation, broadcasting, reproduction on microfilms or in any other physical way, and transmission or information storage and retrieval, electronic adaptation, computer software, or by similar or dissimilar methodology now known or hereafter developed. Exempted from this legal reservation are brief excerpts in connection with reviews or scholarly analysis or material supplied specifically for the purpose of being entered and executed on a computer system, for exclusive use by the purchaser of the work. Duplication of this publication or parts thereof is permitted only under the provisions of the Copyright Law of the Publisher's location, in its current version, and permission for use must always be obtained from Springer. Permissions for use may be obtained through RightsLink at the Copyright Clearance Center. Violations are liable to prosecution under the respective Copyright Law.

Made in the USA
San Bernardino, CA
12 February 2017